North American Mixing Forum

Presentations at the 2008 AIChE Annual Meeting

Philadelphia, Pennsylvania
16 - 21 November 2008

ISBN: 978-1-61567-221-9

Printed from e-media with permission by:

Curran Associates, Inc.
57 Morehouse Lane
Red Hook, NY 12571

CURRAN ASSOCIATES INC. proceedings.com

Some format issues inherent in the e-media version may also appear in this print version.

Printed by Curran Associates, Inc. (2009)

For permission requests, please contact AIChE
at the address below.

AIChE
3 Park Avenue
New York, NY 10016-5991

Phone: (203) 702-7660
Fax: (203) 775-5177

www.aiche.org

Additional copies of this publication are available from:

Curran Associates, Inc.
57 Morehouse Lane
Red Hook, NY 12571 USA
Phone: 845-758-0400
Fax: 845-758-2634
Email: curran@proceedings.com
Web: www.proceedings.com

TABLE OF CONTENTS

37a Some Mixing Issues in the Pharmaceutical Industry

Edward Paul, Self, 308 Brooklyn Blvd., Sea Girt, NJ 08750 and Brian K. Johnson, Merck & Co., Inc., P.O. Box 2000, Rahway, NJ 07065.

Process development and scale-up of processes for the production of active pharmaceutical ingredients require the application of mixing theory and practice for successful operation in manufacturing operations. Mixing issues can be critical components of both homogeneous and heterogeneous reactions either in achieving the desired conversion rate or, of often greater importance, in achieving the predicted selectivity. In addition, many crystallization processes require consideration of mixing in order to realize the desired crystal attributes on scale-up.

Many of the ways mixing issues can impact these operations are discussed along with examples.

37b Effects of Mixing Equipment Configuration and Operation on Fine Chemicals and Pharmaceuticals Manufacture

Victor Atiemo-Obeng, Research & Engineering Sciences, The Dow Chemical Company, 1776 Bldg., Midland, MI 48674 and Arthur W. Etchells III, AWE3 Enterprises, Philadelphia, PA 19106.

Recent developments have opened up the use of ever more novel types of equipment and mixing methods for use in the pharmaceuticals and fine chemicals production. This part of the tutorial will discuss some of the newer developments and put them into context with the more established methods.

37c Fluid Mixing Issues in the Pharmaceutical Industry

Piero M. Armenante, Otto H. York Department of Chemical, Biological, and Pharmaceutical Engineering, New Jersey Institute of Technology, 323 King Blvd., Newark, NJ 07102-1982

Mixing of fluids is an extremely common operation encountered in countless applications in the pharmaceutical industry. Many pharmaceutical processes such as crystallization, precipitation, chemical reaction, fermentation, solid-liquid suspension, and liquid-liquid emulsification, to cite a few, require or are greatly enhanced by the intimate contact between two or more distinct phases (in multiphase systems) or the rapid homogenization of a second fluid (in single phase systems). Despite its ubiquitous presence in pharmaceutical and other industrial processes, mixing is often overlooked in the industrial practice.

This presentation gives an overview of some of the most important fluid mixing issues encountered in the pharmaceutical industry and shows how pharmaceutical processes can benefit from a better understanding of, and a greater appreciation for, the underlying mixing phenomena.

37d The Role of Regulation in Pharmaceutical Production

Otute Akiti, Chemical Engineering and Process Safety, CMC, Exelixis, Inc., P O Box 511, South San Francisco, CA 94083

Regulation of the production of pharmaceuticals plays a big part in the eventual methods used, including the parts of the production involving mixing. Current effects of regulation will be reviewed in this part of the tutorial.

37e Discussion of Mixing for the Production of Pharmaceuticals and Fine Chemicals

Edward Paul, Self, 308 Brooklyn Blvd., Sea Girt, NJ 08750, Victor Atiemo-Obeng, Research & Engineering Sciences, The Dow Chemical Company, 1776 Bldg., Midland, MI 48674, Piero M. Armenante, Otto H. York Department of Chemical, Biological, and Pharmaceutical Engineering, New Jersey Institute of Technology, 323 King Blvd., Newark, NJ 07102-1982, and **Otute Akiti**, Chemical Engineering and Process Safety, CMC, Exelixis, Inc., P O Box 511, South San Francisco, CA 94083.

99a Characterization of Micro-Mixers through Experiments and Multi-Scale Modeling

Christian Lindenberg and Marco Mazzotti. Institute of Process Engineering, ETH Zurich, Zurich, 8092, Switzerland

The attainment of fast mixing is a requirement to avoid transport limitations in processes with fast kinetics such as precipitation at high supersaturations. In this work, the mixing performance of different micro mixers is characterized experimentally and using computational fluid dynamics (CFD).

First, mixing is characterized using a set of competitive-parallel chemical reactions [1]. The mixing efficiency of a wide-angle Y-mixer and a Roughton mixer are compared by measuring the reaction yield at different flow rates, flow rate ratios and reaction time constants [2]. It is found that the Roughton mixer achieves a better mixing performance compared to the Y-mixer and that the mixing efficiency is not affected by the flow rate ratio.

Secondly, the two mixing devices are modeled using computational fluid dynamics (CFD). A closure method using a presumed probability function and the turbulent mixer model is employed to account for micromixing on sub-grid scale [3, 4]. The chemical reaction set is implemented in the CFD code and applied to predict the reaction yield of the mixing experiments which in turn is used to validate the CFD model [2]. Experimental results and model predictions are in good agreement for all mixer geometries and operating conditions. CFD is used to calculate absolute mixing times based on the residence time in the segregated zone and it is shown that mixing times of less than one millisecond can be achieved in the Roughton mixer. CFD provides insight in local concentrations and reaction rates and serves as a valuable tool to improve or to scale-up mixers.

Based on these results the CFD model can be employed for the prediction of the particle size distribution (PSD) in precipitation processes when incorporating the population balance equation (PBE) in the model. However, for most substances nucleation and growth kinetics are unknown, especially at high supersaturations, and the application of a combined CFD-PBE model for real processes might be limited. Therefore, in the third part of this study the mixer setup is used to determine nucleation and growth kinetics of L-asparagine at high supersaturations, i.e. true kinetics which are not affected by transport limitations. The method is based on measuring the PSD obtained at different residence times using a Coulter Multisizer. Nucleation rates are calculated from the change of absolute particle number and growth rates from the change of particle size over time.

References

[1] B.K. Johnson, R.K. Prud'homme, Chemical processing and micromixing in confined impinging jets, AIChE Journal 49 (2003) 2264-2282.

[2] C. Lindenberg, J. Schöll, L. Vicum, J. Brozio, M. Mazzotti, Experimental characterization and multi-scale modeling of mixing in static mixers, accepted for publication on Chem. Eng. Sci. (2008).

[3] J. Baldyga, W. Orciuch, Barium sulphate precipitation in a pipe – an experimental study and CFD modelling, Chem. Eng. Sci. 56 (2001) 2435-2444.

[4] L. Vicum, M. Mazzotti, Multi-scale modeling of a mixing-precipitation process in a semibatch stirred tank, Chem. Eng. Sci. 62 (2007), 3513-3527.

99b Effect of Operating Conditions on the Precipitation of Griseofulvin in Impinging Jets in Stirred-Tank Reactors

Giuseppe Di Benedetto[1], Ankit Patel[1], Ewa Sukcik[1], Piero M. Armenante[1], and Somenath Mitra[2]. (1) Otto H. York Department of Chemical Engineering, New Jersey Institute of Technology, 323 King Blvd., Newark, NJ 07102-1982, (2) Department of Chemistry and Environmental Science, New Jersey Institute of Technology, 323 King Blvd., Newark, NJ 07102-1982

An impinging jet system placed within a stirred-tank reactor was utilized in this work to precipitate an active pharmaceutical ingredient (API), namely griseofulvin, using a solvent/anti-solvent approach. The API was dissolved in an acetone solution. The anti-solvent solution was water added with a polymer (HPMC) and a surfactant (SDS) combination in order to enhance particle stability.

The impinging jet mixer consisted of two jet nozzles arranged diametrically opposed to each other with the outlet tips of the nozzles directed to face each other. The jet nozzles were organized in a submerged arrangement within the stirred-tank reactor. Ultrasonics were also introduced at the point of impingement by way of a sonication probe. The effect of sonication power and angle of impingement of the two jets on the resulting particle size distribution and crystal structure was observed. The particle size distribution was measured for each experiment using a laser diffraction particle size analyzer. The precipitated griseofulvin crystals were also analyzed using Scanning Electron Microscopy (SEM) and X-ray Diffraction (XRD) to obtain an image of the crystal shape and structure.

The impinging jets system produced crystals with smaller mean particle sizes when the two jets were oriented 180 degrees apart and pointed directly at each other. The introduction of ultrasonic power at the impingement point resulted in markedly smaller mean particle size and a tighter particle size distribution. In general, the results were highly reproducible. X-Ray diffraction results showed that the crystal structure was unaffected by different operating conditions.

99c Blend Time in Stirred Tanks Equipped with a High Shear Homogenizer

Micaela Caramellino, Otto H. York Dept. of Chemical Engineering, New Jersey Institute of Technology, 323 King Blvd., Newark, NJ 07102 and Piero M. Armenante, Otto H. York Department of Chemical Engineering, New Jersey Institute of Technology, 323 King Blvd., Newark, NJ 07102-1982.

Blend time is a common measure of the effectiveness of mixing operations in stirred tanks equipped with a centrally mounted impeller. Previous work has shown that blend time is inversely proportional to the impeller speed, but it also depends on geometric and physical parameters, such as the D/T ratio.

Many mixing systems commonly used in the chemical and pharmaceutical industries are equipped with a primary impeller, to promote overall mixing and bulk flow, as well as an additional high-speed rotor-stator homogenizer to generate a high shear zone which enhances shear-sensitive processes, such as emulsification and dispersion, which may be conducted in the tank.

In this work, blending of low-viscosity liquids was studied in fully baffled stirred tanks equipped with a Rushton turbine as the primary impeller and a high-speed homogenizer. Blend time was experimentally determined using an acid-base de-colorization technique coupled with an imaging analysis technique, as well as through conductivity measurements. Experiments were first conducted with the primary impeller operating alone, and then with both the primary impeller and the homogenizer in operation. Different combinations of the agitation speed of the primary impeller and the homogenizer speed were investigated. The data show that the additional mixing effects generated by the homogenizer produces a significant reduction in the overall blend time obtained with the primary impeller alone, especially when feeding in the homogenizer zone. A semi-empirical model was used to analyze and interpret the experimental data.

99d Monitoring Dissolution of High Viscosity Liquids In Water

Thomas L. Rodgers[1], Michael Cooke[1], Flor R. Siperstein[1], and Adam J. Kowalski[2]. (1) School of Chemical Engineering and Analytical Science, University of Manchester, Po Box 88, Sackville Street, Manchester, M60 1QD, United Kingdom, (2) Unilever R&D Port Sunlight, Quarry Road East, Bebington, Wirral, CH63 3JW, United Kingdom

Emulsions are often found in formulated products. They contain small liquid droplets dispersed into an immiscible liquid that remains as a continuous phase. Emulsions are not thermodynamically stable systems; therefore their properties depend strongly on the preparation method and material history. During the preparation of emulsions, it is often desirable to dissolve highly viscous fluids, such as surfactants or other additives, into an aqueous phase. In this work we measure dissolution times for a non-Newtonian highly viscous surfactant (sodium laureth sulphate) in water using vessels of different diameter ($D = 0.9$ and $D = 0.15$ m) with high shear disk agitators ($D/T = 1/3$). To monitor the dissolution times we use a single probe conductivity measurement and volume averaged conductivity obtained by 3D electric resistance tomography. We compare the validity of the methods and highlight the difference between dissolution times for highly viscous fluids and mixing times for tracers with the same viscosity as water.

99e Prediction of Velocity Field and Particle Deformation In An In-Line Rotor-Stator Mixer

B. N. Murthy[1], Karl R. Kevala[1], Kenneth T. Kiger[2], and Richard V. Calabrese[1]. (1) Chemical & Biomolecular Engineering, University of Maryland, Building 090, Room 2113, College Park, MD 20742-2111, (2) Mechanical Engineering, University of Maryland, College Park, MD 20742-3035

In-line rotor-stator mixers are employed to produce and control particle size in liquid-liquid and solid-liquid dispersions. With respect to ultimate particle size, these devices often operate far from equilibrium and fluids passing through them experience a limited residence time in the highest shear regions of the flow field. Models that quantify breakage phenomena in the kinetic regime are quite limited and a fundarnenta1 basis for scale-up of in-line rotor-stator, multiphase processes is lacking.

The ultimate goal of this study is to quantify dispersion and emulsification processes in in-line rotor-stator mixers operating in the turbulent regime. Previously, we reported results using a fast particle tracking algorithm in conjunction with RANS sliding mesh CFD simulations, for a Silverson L4RT in-line device. We demonstrated that we could obtain physically realistic results for the paths of particles of various density and size ('real particles'). We have now further developed our simulation techniques to track 'real particles' to acquire their shear rate history and to statistically analyze particle trajectories to provide practical suggestions for scale-up. Furthermore, we have preformed parametric studies varying operating conditions (rotor speed and throughput) and particle properties.

We will report systematic results for the flow field and particle path/shear rate history for several operating conditions. In addition to qualitative comparisons, statistical data will be provided to quantify the effect of operating conditions and particle properties on shear history and particle residence time. Our ultimate goal is to further develop the particle tracking code into a population balance algorithm to allow prediction of particle size in liquid-liquid and solid-liquid systems. We will discuss our future plans in the context of what can realistically be accomplished in the short term.

99f Studies on Mass Transfer on Inner Surface of An Annular Column with Coaxially Placed Spiral Coil as Turbulent Promoter

Ganapathy Y, **P. Rajendra Prasad**, Vaka Murali Mohan, **Sujatha V**, and Sarveswara Rao S. Chemical Engineering, Andhra Universirty, Visakhapatnam, 530003, India

Studies on mass transfer on inner surface of an annular column with coaxially placed spiral coil as turbulent promoter was presented in this paper. The electrolyte is 0.01 M equimolal Potassium ferricyanide, Potassium Ferro cyanide and 0.5N sodium hydroxide. The variables covered in this study are the flow rate of the electrolyte, the geometric parameters of the promoter. The pitches of the coil covered in the study are 0.01 m, 0.02 and 0.03m. The diameter of the coil in the column is varied as 0.02 m, 0.03 m and 0.40 m. Reynolds number covered in the study ranges from 1000 to 15000. Enhancements in friction factors are 11 folds over the flow without promoter in the column. A model is developed in terms of geometric parameters and presented.

158a Micromixing In Capillary Channels with Buoyancy Flows and Reaction at the Walls. Effect of Asymmetric Conditions

Katherine Cerda[1], **Mario Oyanader**[2], and Pedro E. Arce[2]. (1) Chemical Engineering, University Catolica del Norte, Avda Angamos 0610, Antofagasta, Chile, (2) Chemical Engineering, Tennessee Technological University, Department of Chemical Engineering, P.O. Box 5013, Cookeville, TN 38505

Micromixing in Capillary Channels is a important aspect that applies to different processes, such as electro-assisted drug delivery, micro-electrophoretic separations, soil remediation among others. In these cases, the hydrodynamics of a system can considerably change the dispersion of solute inside a capillary channel, due effect of buoyancy driven type flows on effective parameters, i.e., effective dispersion coefficient and effective convective velocity. This contribution includes a reactive surface at the walls (first order) to analyze retardation effects. In this approach, explicit analytical expressions have been derived for the effective parameters as a function of the ambient temperature, skew temperature parameter and net mass flow. The spatial averaging method, in combination with the solute species continuity equation, has been use to determine the effect of buoyancy driven type flows on "effective parameters" and to predict the macroscopic behavior of the system in a rectangular geometry. Several illustrations are presented to demonstrate the "macromixing" and retardation effect caused by buoyancy and wall reaction respectively. Specific cases of effective dispersion coefficient and effective convective velocity trends as function of the Thiele modulus, Grashoff and Schmidt numbers are analyzed.

158b Microfluidic Manipulation of Mass Transport In Nanoparticle Suspensions

Serdar Ozturk[1], Yassin A. Hassan[2], and Victor M. Ugaz[1]. (1) Artie McFerrin Department of Chemical Engineering, Texas A&M University, 3122 TAMU, College Station, TX 77843, (2) Departments of Nuclear and Mechanical Engineering, Texas A&M University, 3122 TAMU, College Station, TX 77843

Low volume-fraction (~1%) suspensions of solid nanoparticles (1–100 nm diameter) have been observed to exhibit remarkably enhanced heat transport properties relative to the particle-free fluid alone. These unique characteristics have generated intense interest in harnessing these so-called nanofluid suspensions for use in thermal management applications. Although most work to date has focused on enhancement of thermal properties, recent studies have also suggested that a corresponding increase in effective mass diffusivity is achievable in aqueous suspensions of 20 nm alumina (Al_2O_3) nanoparticles relative to that of pure water. Although this raises exciting possibilities for manipulation of microscale mixing, there are still insufficient data and theoretical insights to draw meaningful conclusions about the magnitude and scalability of these effects.

Here we employ a microfluidic system that allows us to investigate this phenomena by direct visualization of fluorescent tracer diffusion between nanoparticle-laden fluid streams as they travel through a microchannel network. This is accomplished using both "top view" imaging of parallel dye streams and confocal laser scanning microscopy to obtain a cross-sectional view of the flow. We investigate aqueous suspensions of Al_2O_3 nanoparticles with mean size ranging from 20 to 100 nm at concentrations ranging from 0.001 to 1 vol%. Diffusion coefficient values associated with the tracer dye were determined by measuring the growth of the mixed interfacial zone between the streams. Suspension viscosity and thermal conductivity were also characterized, and particle size was determined using dynamic light scattering. A key observation of this work is that the observed dye transport is strongly dependent on the composition of the nanoparticle suspension and that these parameters must be carefully controlled in order to draw correct conclusions about diffusivity. For example, we observe that the measured diffusivity of a food dye tracer nearly doubles as the nanoparticle concentration is increased to 0.5 vol% while a much more modest change is observed with Rhodamine B. These effects arise as a consequence of interactions among the nanoparticles, dye molecules, and stabilizing additives. Understanding the nature of these interactions has the potential to enable new ways to manipulate mass transport and mixing in microfluidic systems.

158c Dispersion Effects of Microchannel Configurations on Microgc Performance

Adarsh D. Radadia, Chemical and Biomolecular Engineering, University of Illinois, 600 S. Mathews St., Urbana, IL 61801, Rich I. Masel, Chemical and Biomolecular Engineering, University of Illinois at Urbana-Champaign, 294 RAL, MC-712, 600 S. Mathews Ave., Urbana, IL 61801, and Mark A. Shannon, Mechanical Science and Engineering, University of Illinois, 600 S. Mathews Ave, RAL 204, Urbana, IL 61802.

Microchannels are essential components of microreactor or microseparation devices. In the process of accommodating long microchannels on a small chip area, there exists a choice of channel configurations. The objective of this paper is to understand how microchannel configurations the separation performance of a micro gas chromatographic column operating under laminar conditions.

Here we present results of extensive tests done using three different configurations: 1)Serpentines, 2) Circular Spirals and, 3)Square Spirals. The results presented here will help future microGC designers in designing high separation efficiency columns. The microchannels are fabricated in silicon and connected to a conventional gas chromatograph operating in high speed mode. The paper will present RTD measurements to compare the gas chromatographic performance of different channel configurations. Effect of configurations is demonstrated on separation of analytical chemical mixtures.

HYDRODYNAMIC CONDITIONS IN A SIMPLE ROTATING MICROBIOREACTOR INDUCE CHANGES IN EXCRETION LEVEL AND GLYCOSYLATION PATTERNS OF THE Von WILLEBRAND FACTOR (vWF) IN ENDOTELLIAL CELLS

Morga-Ramírez M., Collados-Larumbe, M.T., Johnson, K., Rivas M.J., Carrillo-Cocoom, L.M., and M.M. Álvarez[ϒ]

Centro de Biotecnología. Instituto Tecnológico y de Estudios Superiores de Monterrey. Ave. Eugenio Garza Sada 2501 sur. Monterrey, N.L. México. C.P. 64849.

Abstract.

There is an ongoing debate on the extent of the effect of mechanical stress (if any) on the performance of mammalian cell cultures. Here, we study the effect of low speed rotational agitation on the level of expression and the glycosylation patterns of the Von Willebrand factor in endothelial cells cultured in 6 cm Petri dishes placed at different locations on a rotating platform. In this simple system, complex and non-intuitive velocity fields arise. Flow behavior strongly varies depending on radial distance and rotational velocity value, as revealed by simple calculations of tangential velocities and CFD analysis. We found significant differences in both, excretion level and glycosylation patterns (extend and completeness of glycosylation), as a result of relatively small changes in rotational speed. Our results support the hypothesis that differences in mixing conditions may induce important changes in the secretion and assembly of glycoproteins in mammalian cell cultures.

Key words: Endothelial cells, mixing, vWF, shear stress, glycosylation, secretion.

INTRODUCTION

In this contribution, we report significant changes on the secretion of Von Willebrand Factor (vWF) and in two particular indicators of glycosylation (trimannose and sialic acid concentration) when endothelial cells are exposed to incremental rotational fields in Petri dishes. The vWF is a well-characterized plasma circulating glycoprotein produced and secreted particularly by endothelial cells (see for example Vischer 2006). vWF plays essential roles in hemostasis. It mediates platelet adhesion to the vascular wall, platelet aggregation, and serves as a plasma carrier for factor VIII, stabilizing it in the circulation. Also, it has been proposed as a marker for prognosis in primary pulmonary hypertension patients (Collados *et al.*, 1999), thrombosis and cardiovascular disease (Vischer, 2006). The over-secretion of vWF

[ϒ] Corresponding autor: Mario.Alvarez@itesm.mx

has been associated to the complex biochemical cascade implicated in thrombotic events (Galbucera *et al.*, 1997). The role of vWF in platelet adhesion is particularly important under conditions of high shear stress.

Indeed, it is well known that certain mammalian cells (i.e. endothelial vascular cells) are particularly responsive to mechanical stresses (Dangaria and Butler 2007, Galbucera *et al.*, 1997). Typical local velocity values of arterial flows promote alignment to the flow field and secretion of vasodilatant and anti-coagulant compounds (Macintire *et al.*, 1998). However, when abnormal flow conditions occur in arterial flow, endothelial cells might also produce substances that promote vasoconstriction, coagulation and platelet aggregation (Sokabe *et al.*, 2004). Indeed, some crucial metabolic processes related to the coagulant cascade or cardiovascular disorders are triggered or modulated by the hydrodynamics of blood in veins and arteries (Chiu *et al.*, 2005).

MATERIAL AND METHODS

Isolation, cell culture and propagation of endothelial cells. Human Umbilical Vein Endothelial Cells (HUVEC) were isolated from umbilical human cordons following protocols detailed elsewhere (Rodríguez *et al.*, 2002; Morga-Ramírez, 2007). Cells were recovered by incubation of the umbilical cordon vein with HSS-0.2% colagenase type II (Gibco BRL) for 15-20 minutes at of 37°C. Cells were centrifuged, washed, and re-suspended in supplemented culture media M199 [20% FBS (Gibco 12318-020), 5 U/ml heparine, 1% of antibiotics solution (penicillin and streptomycin), 10 mM HEPES, 2 mM glutamine, and 0.40 mg/ml of Endothelial Growth Factor]. Cells were seeded in culture bottles of 25 cm^2 (Corning) in 4ml of supplemented M199 culture media (previously described) and incubated at 37°C and 95% relative humidity in a 7% CO_2 and 93% air atmosphere.

Experimental system. Cells were culture in polystyrene Petri dishes (6 cm diameter) from Corning®, U.S.A. Petri dishes were placed in a rotating platform of 25 cm in diameter, at various radial distances with respect to the axes of rotation (see figure 1). Confluent cultures, normally obtained after 7 days of culture, and changing culture media every second day, were obtained in static conditions (0 RPM). Afterwards, cultures were exposed to different rotational laminar velocity fields. Here, we report results from experiments conducted at 0, 3, 5, and 8 RPM. Levels of secreted vWF after 24 hours of exposition to rotational fields were evaluated using an immunoassay technique. In addition, extent of glycosylation was evaluated by measuring the amount of trimannose incorporated to the vWF secreted at each experiment (technique described below). Level of glycosylation completeness was evaluated by measuring the amount of sialic acid incorporated to the secreted vWF in each experiment (technique described below).

Flow characterization using CFD. Computational fluid dynamics (CFD) was used to characterize flow behavior within Petri dishes rotating at different velocities. A solid model of the initial fluid volume of the Petri dish was made using the parasolids capability of Synergy, from Weber Systems Inc., Menomonee Falls, WI. The solid model was imported into the solver environment used in this work (ACUSOLVE CFD) using the acuConsole program also from ACUSIM Software Inc.

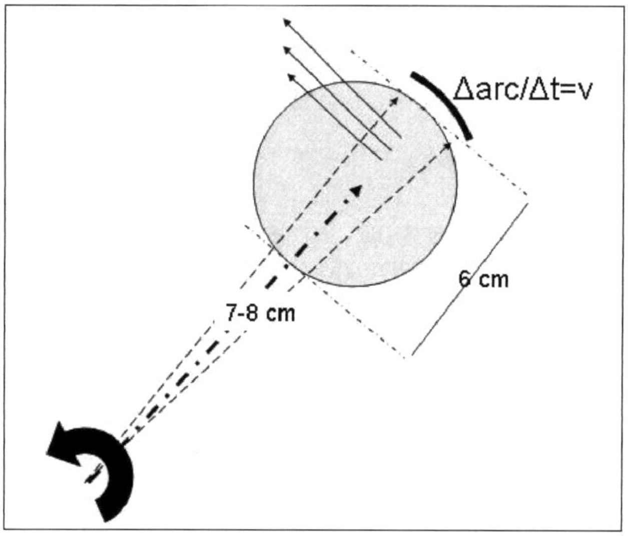

Figure 1. Endothelial cells were culture in Petri dishes of 6 cm in diameter, located in a rotating platform of 24 cm diameter; within the platform, local values of the tangential velocity augments as radial distance increases.

The mesh was generated using the acuMeshSim program, which also facilitated the generation of very refined boundary layer elements from the dish surface. The resultant model consisted of over 755,000 tetrahedra elements, and 137,000 nodes. The CFD solver used in this work was ACUSOLVE CFD from ACUSIM Sofware Inc., Mountview Ca., code based on a special stabilized Galerkin/Least Squares (GLS) finite element technology formulation of the full transient 3D non-linear Navier-Stokes equations of fluid dynamics. The fluid was modeled primarily as a newtonian fluid with the nominal properties of water.

Protein secretion assays. Released vWF antigen was measured in samples taken from the culture supernatant of the Petri dishes using a sandwich immunoenzymatic assay detailed by Morga-Ramírez (2007).

Glycosylation assays. In order to investigate the effect of low Re rotational agitation on the glycosylation patterns of vWF, we measured the total concentration of trimannose and terminal sialic acid resides on vWF released by confluent cultures of endothelial cells agitated at different rotational speeds. Mannose and sialic acid content was calculated by an enzyme lynked lectin assay (ELLA) slightly modified from that reported by Collados *et al.* (1999).

RESULTS AND DISCUSION

Analysis of local velocities. Although geometrically simple, and laminar in nature, the flow system analyzed here is highly complex. Interestingly, a wide range of local velocities can be imposed to the cultured cells by varying the rotational speed of the platform or the radial distance at which the center of the Petri dish is positioned. Cells located in the vicinity of the platform perimeter experience local speeds as high as 5 m/s (8.38 cm/s) when rotated at 8

RPM. Lower local velocity values are observable as we approach the center of the rotating platform or as the rotational velocity decreases. This range of local velocities translates into a range of Reynolds number values (Re) from 0 to 100, still well into the laminar regime. Also, Γ_{xy}, the maximum normal shear stress exerted by the inertial momentum of the culture media over the layer of cells growing at the bottom surface of the Petri dishes, can be estimated by the equation: Γ_{xy} (dynas/cm^2)= $-\mu$ dv_x/dy. In this equation, v_x is the velocity component tangential to the cell layer and y is the distance in the direction normal to the flow. The value (dv_x/dy) is the strain rate due to the velocity component in the x direction (tangential to the bottom of the Petri dish). Consequently, the shear stress felt by the monolayer of endothelial cells growing at the bottom of the Petri dish (consider that the cell monolayer is 10 microns height), ranges from 60 to 0 dynes/cm^2. A more comprehensive and realistic characterization of the 3D velocity field, the flow patterns, and the shear stresses induced by the platform rotation inside the Petri dishes is feasible using CFD (Computational Fluid Dynamics) analysis.

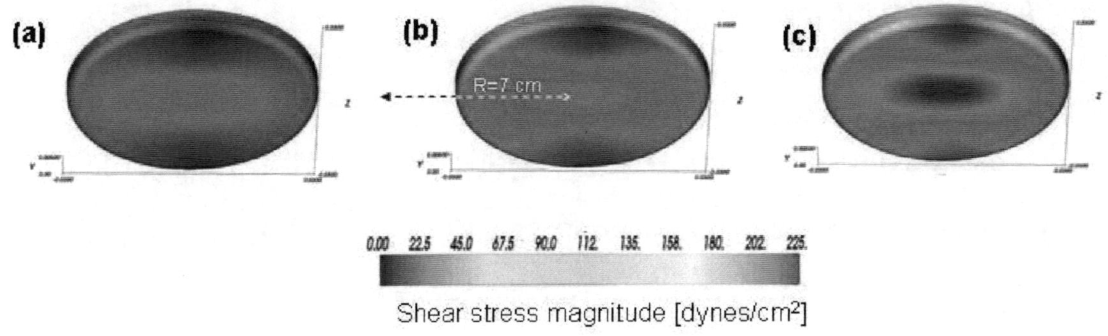

Figure 2. Shear stress values calculated by CFD at 0.05 cm from the bottom surface of Petri dishes rotating at (a) 3 RPM, (b) 5 RPM, and (c) 8 RPM.

Figure 2 presents CFD calculations of the strain rate fields prevalent just above the layer of anchored cells. For a Newtonian system, the shear stress is proportional to the shear rate, being the corresponding proportionality constant the fluid viscosity. If we consider that v_y•0 (equivalent to state that the vertical velocity component is very small compared to the radial and tangential velocity component) then (Γ_{xy}) at each point of the cell monolayer can be approximated by the product of the viscosity times the strain rate calculated by CFD. Moreover, since the typical range of viscosities of mammalian cell culture media approximately averages 1 cp, shear stress values are approximately equivalent to the strain rate values. The highest shear zone is again located at the central areas of each Petri dish. However, CFD analysis reveals very interesting, heterogeneous and non expected shear stress spatial distributions. Globally, maximum shear rates above 50 dynes/cm^2 occur at the central area of dishes rotated at 8 RPM. In the culture systems rotating at 3 and 5 RPM, maximum shear rates do not exceed 40 dynes/cm^2. Subsequently, we will review the effects of these complex flow and shear stress fields in the level of secretion and glycosylation extent of HUVEC cells under culture.

Effect of agitation on von Willebrand secretion levels. Galbucera *et al.* (1997) grew human endothelial cells in a cone and plate system, a well studied flow device, under laminar flow conditions. In these experiments, the cell monolayer was exposed to shear stress values also from 0-12 dynes/cm^2. They found that the secretion of the von Willebrand factor was modulated by the shear stress. As shear stress augments, the von Willebrand concentrations in the extra-cellular environment increased proportionally. In our experiments, we explored a wider range of shear stress values (0-60 dynes/cm^2). In average, rotational speeds as low as 3 RPM (Re•30; Γ_{xy}•30.0 dynes/cm^2) stimulate cells to increase the release of vWF concentration 2.44 fold with respect to non agitated conditions. At 5 RRM (Re•50; Γ_{xy}•37 dynes/cm^2) a statistically similar secretion value was observed with respect to 3 RPM experiments. At 8 RPM (Re•80.0; Γ_{xy}•55.0 dynes/cm^2), the excreted vWF concentration was, in average, 3.21 times that observed in static experiments. Our results are consistent with those reported by Galbucera *et al.*, (1997) for the low Re regime (Re•30; Γ_{xy}•30.0 dynes/cm^2). However, for the range of Re>50 (ranged not explored before) we found progressively lower vWF secretion levels, as the rotational speed increases.

Effect of agitation on Von Willebrand glycosylation. Glycosylation is a highly complex cellular process, dependant on the 3-dimensional structure of the protein, the enzyme repertoire of the host cell, the transit time in the Golgi and the availability of intracellular sugar-nucleotide donors (Butler, 2006). A fundamental objective of this communication is to investigate the effect of different rotational protocols on the extent of glycosylation of the vWF secreted by HUEVEC cultures. As elegantly demonstrated by Naismith and Field (1996), Concanavalin A binds trimannosides, specifically the trimannosides core for biantenary N-linked glycans (see figure 3).

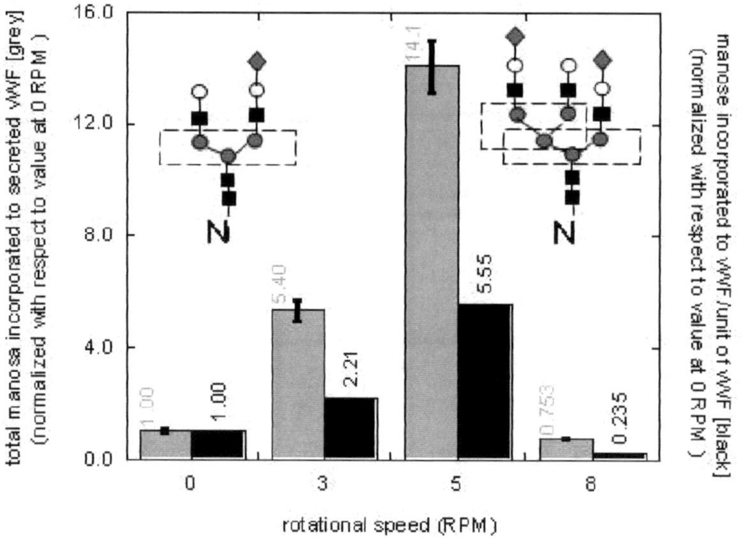

Figure 3. Glycosylation extent at different rotational conditions. Evaluation done by determination of the amount of trimannose core structures using Concanavalin A. Data has been normalized with respect to secretion at static conditions. Each bar presents results from 3 to 4 replicas. Glycans presented in the figure depict location of trimannose cores (within dashed rectangle), common structures to all N-glycan forms.

The fore mentioned authors crystallized the Con A-trimmanoside complex, identifying the Con A binding site and calculating the distances between each one of the three mannoses of the glycan core and the lectin aminoacid residues responsible of the binding specificity. Due to this specificity, Con A binding extent can be an adequate indicator for N-glycosylation extent, since the trimannose core is a common structure among both clases of N-glycans, high-mannose and complex glycans (see for example Manzella *et al.*, 1996).

Figure 3 presents experiments at which Con A is added to confluent HUEVEC cultures to measure glycosylation extend of the secreted vWF. Results have been normalized with respect to the values registered at static conditions (light gray bars). The overall trimannose concentration, as determined by our Con A assay, was dramatically different in Petri dishes rotated at different speeds. Overall trimannose concentration was, in average, 5.40 times higher in dishes rotated at 3 RPM than in static experiments. In 5 RPM experiments, overall trimannose concentration was almost three fold higher than observed at 3 RPM and 14.1 fold that observed in static conditions. Not intuitively, trimannose concentration in Petri dishes rotated at 8 RPM dropped significantly, 25% lower with respect to 0 RPM experiments. To account for different expression levels at each agitation condition, a second normalization, this time with respect to the vWF excreted, is presented (dark gray bars). The extent of glycosylation per molecule also varies significantly between agitated and non-agitated conditions. Remarkably, the ratio of trimannose/vWF molecule secreted in experiments at 3 and 5 RPM is, respectively, 2.21 and 5.55 fold higher than that observed in static cultures. This result made evident that agitation, even at low Re values (•30) can promote dramatic changes in glycosylation activity. Interestingly, in the experiments at 8 RPM, the extent of gliycosylation is significantly lower than that observed in experiments at 5, 3 and 0 RPM (trimannose/vWF= 0.235) making explicit a non-linear and significant dependence of glycosylation extent with respect to agitation.

In addition, we conducted experiments to measure the sialic acid content in vWF as a second indicator of glycosylation quality. Since sialylation is the last intracellular stage of the glycosylation process that takes place in the Golgi, its extent is an indicator of terminal quality of the glycosylation process (see for example Chuan *et al.*, 2006; Shauer 2000, and Gramer *et al.*, 1995). Complete glycosylation of recombinant proteins is usually associated with maximization of galactosylation and sialylation (Butler, 2006). Figure 4 presents sialic acid content of vWF secreted in the supernatant in confluent HUVEC cultures exposed to different rotational velocities, as determined through binding by SNA (a lectin that exhibits high specificity for sialic acid terminal structures). Results are normalized with respect to the values observed for static cultures. Note that the average content of sialic acid determined in experiments at 3 and 5 RPM is, respectively, 6.33 and 7.64 times higher than that observed for static cultures. However, in experiments at 8 RPM, the overall sialic acid content was only 1.40 times that observed in static cultures. If a second normalization is done, now considering that the amount of vWF excreted at different rotational conditions, results from experiments at 3 and 5 RPM are statistically similar (exceeding approximately three fold the values measured for non agitated conditions). Interestingly, in the experiments at 8 RPM, once normalized by the amount of vWF excreted in those experiments, the sialic acid/vWF ratio is lower than that observed in static experiments. At normal conditions, vWF is a highly sialylated glycoprotein. Approximately 56% of their N-linked oligosaccharides are monosialylated and 30% are disialylated (Matsui *et al.*, 1992). Consequently, in average, per glycan molecule linked to vWF,

1.06 sialic acid terminals are present. From an analogous analysis, there are 1.09 trimannose core structures by glycan molecule. Consequently, the expected trimmanose/sialic acid ratio in vWF is 1.03. Our results indicate trimannose/sialic acid ratios of 0.85 and 1.84 for our experiments at 3 and 5 RPM respectively. Therefore, the significant increase in glycosylation extent observed at rotational values of 5 RPM is not accompanied by a proportional sialylation.

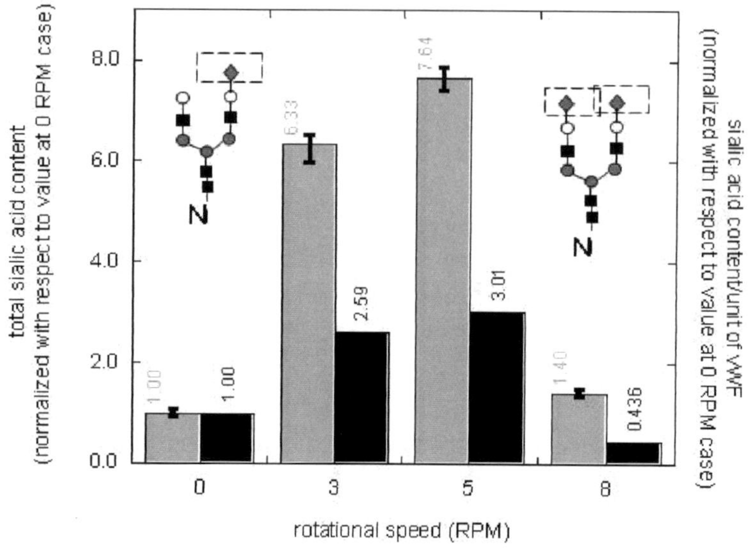

Figure 4. Glycosylation completeness at different rotational conditions, as evaluated by determination of the amount of sialic acid using SNA. Data has been normalized with respect to secretion at static conditions. Each bar presents results from 3 to 5 replicas. Glycans presented in the figure depict location of sialic acid terminal residues (within dashed rectangle). Each glycan might have or not sialic acid residues terminating each antenna.

CONCLUSIONS

Rotational speeds in the low Reynolds regime have significant effect on the levels of secretion and the glycosylation patterns of the von Willebrand factor expressed by HUVEC cultures. This contribution confirms the amplified secretion of vWF observed before in endothelial cell cultures exposed to moderate shear stress levels. Additionally, we document, to our knowledge by the first time, significant changes in glycosylation patterns in the vWF under different shear stress and/or agitation conditions. Our results support the hypothesis that even modest differences in agitation conditions may induce important changes in the secretion and assembly of glycoproteins in mammalian cell cultures.

Our observations might also have relevant implications in the medical context, particularly on the understanding of the complex biochemical cascade related to cardiovascular disorders..

REFERENCES

1. Butler M. 2006. Optimization of the cellular metabolism of glycosylation for recombinant proteins produced by mammalian cell systems. *Cytotechnology* 50(1-3): 57-76.
2. Chiu JJ, Chang SF, Lee PL, Lee CI, Tsai MC, Lee DY, Hsieh HP, Usami S, Chien S. 2005. Shear stress inhibits smooth muscle cell-induced inflammatory gene expression in endothelial cells- Role of NF-kappa B.*Arteriosclerosis Thrombosis and Vascular Biology* 25(5):963-969.
3. Chuan KH, Lim SF, Martin L, Yun CY, Loh SOH, Lasne F, Song ZW. 2006. Caspase activation, sialidase release and changes in sialylation pattern of recombinant human erythropoietin produced by CHO cells in batch and fed-batch cultures. *Cytotechnology* 51(2): 67-79.
4. Collados MT, Sandoval J, López S, Massó FA, Páez A, Borbolla JR and Montaño LF. 1999. Characterization of von Willebrand factor in primary pulmonary hypertension. *Heart Vessels* 14:246-252.
5. Dangaria JH and Butler PJ. 2007. Macrorheology and adaptive microrheology of endothelial cells subjected to fluid shear stress. *American Journal Of Physiology-Cell Physiology* 293(5):C1568-C1575.
6. Galbusera M, Zoja C, Donadelli R, Paris S, Morigi M, Benigni A, Figliuzzi M, Remuzzi G, Remuzzi A. 1997. Fluid shear stress modulates von Willebrand factor release from human vascular endothelium. *Blood* 90(4): 1558-1564.
7. Gramer MJ, Goochee ChF, Chock VY, Brousseau DT and Sliwkowski MB. 1995. Removal of sialic acid from a glycoprotein in cho cell culture supernatant by action of an extracellular cho cell sialidase. *Nature Biotechnology* 13, 692 – 698.
8. Macintire LV, Wagner JE, Papadaki M, Whitson PA, ans Eskin SG. (1998). Effect of Flow on Gene Regulation in Smooth Muscle Cells and Macromolecular Transport Across Endothelial Cell Monolayers. *The Biological Bulletin* 194(3):394-399.
9. Manzella SM, Hooper LV, and Baenzinger JU. (1996). Oligosaccharides containing b1,4-linked N-Acetylgalactosamine, a paradigm for protein-specific Glycosylation. *The Journal of Biological Chemistry* 271(21):12117-12120.
10. Morga-Ramírez M. (2007). Efecto del Estrés mecánico o de Concentraciones Elevadas de Homocisteina sobre los Patrones de Glicosilación del Factor von Willebrand en Células Endoteliales *in vitro. Master Degree Thesis.* Tecnológico de Monterrey. México. 93 pp.
11. Naismith J.H. and Field RA. (1996). Structural Basis of Trimannoside Recognition by Concanavalin A. *The Journal of Biological Chemistry* 271(2): 972-976.
12. Shauer R. 2000. Achievements and challenges of sialic acid research. *Glycoconjugate Journal* 17(7-9): 485-499.
13. Sokabe T, Yamamoto K, Ohura N, Nakatsuka H, Qin K, Obi S, Kamiya A and Ando J. (2004). Differential regulation of urokinase –type plasminogen activator expression by fluid shear stress in human coronary endothelial cells. *Am J Physiol Heart Circ Physiol.* 287(5): H2027-H2034.
14. Vischer U.M. 2006. von Willebrand factor, endothelial dysfunction, and cardiovascular disease. *Journal of Thrombosis and Haemostasis* 4: 1186–1193.

158e a Numerical Study of Induced Charge Electro-Osmotic Mixing

Mranal Jain, A. Yeung, and K. Nandakumar. University of Alberta, 536 Chemical and Materials Engineering Building, Edmonton, AB T6G 2G6, Canada

Mixers used in the microdevices can be categorized as active or passive. Active mixers utilize external energy (i.e. pressure, electrokinetic disturbance etc.) for mixing and inducing transverse flow. On the other hand, diffusion or chaotic advection is the dominant mixing mechanism in passive mixers. The most basic passive micro-mixer is a T-mixer or Y-mixer, where two different solutions flow adjacently in the same direction and mixing is primarily due to diffusion in the transverse direction. There are various modifications reported in the literature for enhancing mixing performance for T-mixers. Some of reported approaches include sequential injection of samples, patterned heterogeneous surface charge along the walls, grooved patterns on the channel base, etc.

In this numerical study, we present a novel micro-mixer design which utilizes the recent concept of induced charge electro-osmosis (ICEO) for enhancing mixing performance in a T-mixer. Any external electric field will induce charges on the metal surface which in turn leads to the formation of an electric double layer (EDL) around the metal surface. The electric field will then act on the diffuse part of the EDL to generate different kinds of flow pattern within the micro-channel. In this particular case, micro-vortices can be generated using ICEO, which enhances the mixing performance in the device. The proposed mixer design is compared with the traditional mixer design and found to be superior in terms of performance. The numerical results are found to be in agreement with analytical results available for electrokinetic flow and electrokinetic mixing for the conventional T-mixer. Further, the effect of design and operational parameters on the mixing performance is analyzed for the new design.

158f Numerical Simulations of Bubble Formation and Rise in Micro-Channels

Deepak Goel and **Vivek V. Buwa**. Chemical Engineering, Indian Institute of Technology-Delhi, Hauz Khas, New Delhi, 110016, India

The gas-liquid flow in micro-channels is of fundamental importance for many engineering applications like micro-reactors, monolith reactors, micro-mixers and heat exchangers and several other micro-fluidic devices. In this work, the bubble formation and their rise in circular capillaries in Taylor flow regime is investigated by using the Volume of Fluid (VOF) method. The dynamics of formation and rise of Taylor bubbles in glass capillaries of 1 mm, 0.5 mm, 0.75 mm and 0.3 mm diameter for air-water and air-octane systems was simulated. The effects of superficial gas and liquid velocities, channel geometry (nozzle wall thickness, nozzle diameter, capillary diameter), wall adhesion (contact angle), and fluid properties (surface tension, viscosity) on the dynamics of bubble formation was investigated. The predicted bubble shapes and bubble formation periods were validated using the experimental data reported by Salman et al. (2006) for a wide range of experimental parameters.

The key conclusions of our investigations are summarized here:

• The dynamics of bubble spreading and receding on the nozzle wall is important and therefore it is necessary to consider the nozzle geometry precisely in the simulations.

• At coarse grid resolutions, predictions of axi-symmetric simulations predictions were found to be in good agreement with experiments, but the predicted results were not grid independent and lead to unrealistic results as the grid was refined. Therefore, axi-symmetric simulations are not adequate to investigate the bubble formation mechanism in micro-channels and 3-D simulations are required for accurate prediction of bubble formation mechanisms in micro channels.

• The predicted bubble formation periods were in good agreement (within 10%) with the measurement for UL> 0.012 m/s and UG < 0.02 m/s. However, for UL < 0.0075 m/s and UG > 0.02 m/s, bubble modification due to bubble pairing and coalescence at the nozzle was observed and deviation of numerical predictions from experimental data increased, though the experimental data in itself contains large deviations (as large as 50 %) in such cases.

• The wall adhesion was found to strongly influence the bubble formation mechanism and must be considered in the simulations for accurate prediction of bubble formation dynamics. Bubble formation period was found to decrease with the increase in the capillary wall contact angle and to increase with the increase in the nozzle wall contact angle.

• The predicted bubble formation periods were found to strongly depend on the surface tension up to 0.072 N/m. With the increase in the surface tension, bubble tends to become more spherical, the bubble diameter and hence the bubble formation period was to found increase. For the range of liquid viscosities considered in the present work (0.0001 to 0.01 Pa.s), the effect of viscosity on bubble formation was found to be negligible which agreed well with the previous investigations reported in the literature.

• Bubble formation is sensitive to gas and liquid inlet geometry and channel dimensions. The bubble formation period was found to decrease with the decrease in the nozzle diameter and the capillary diameter.

The extensive validation of the computed bubble shapes and formation periods with the experimental results demonstrate that the Taylor flow regime in micro-channels can be successfully predicted for wide a range of operating parameters. The experimentally validated computational model will be very useful to simulate the transport processes and reactions in micro-capillaries/channels.

Reference:

Salman, W., Gavriilidis, A., Angeli, P. On the formation of Taylor bubbles in small tubes. Chemical Engineering Science. 2006, 61, 6653 – 6666.

248a Mixing on the Microscale: Interaction of Stoichiometry and Reaction Rate with the Batchelor Time Scale

Syed Imran A. Shah[1], Larry W. Kostiuk[2], and **Suzanne M. Kresta**[1]. (1) Chemical and Materials Engineering, University of Alberta, 536 Chemical and Materials Engineering Building, Edmonton, AB T6G 2G6, Canada, (2) Mechanical Engineering, University of Alberta, 4-9 Mechanical Engineering Building, Edmonton, AB T6G 2G8, Canada

When a reaction is nearly instantaneous it is often mixing limited, i.e. the product yield is dependent on how fast the reactants are brought together. This effect of mixing is especially evident in the case of a competitive-consecutive reaction scheme where the product of the desirable reaction reacts with one of the reactants to form undesirable by-products. This kind of reaction mechanism can result in large yields of undesirable by-products, sometimes up to two or three times the yield of the desired product. A 1-D, non-dimensional, diffusion-reaction model at the micro-mixing scale has been developed to attempt to characterize this reactive-diffusive system. Constant density and diffusivities were assumed and the system of PDE's was solved using COMSOL 3.4. It was found that the ratio of rate constants of the reactions, the stoichiometry of the reactions, and the mixing timescale have considerable effects on the yield of desirable product.

248b An Investigation of the Superblend Mixer

Maya Farhat[1], Louis Fradette[1], Philippe A. Tanguy[1], Hiro Horiguchi[2], and Ryuichi Yatomi[2]. (1) URPEI, Chemical Engineering Department, Ecole Polytechnique de Montreal, P.O. Box 6079, Station Centre-Ville, Montreal, QC H3C3A7, Canada, (2) R&D Dept., SHI Mechanical & Equipment Inc., 1501 Imazaike, Toyo City, Ehime, Japan

Coaxial mixers are known to be well adapted to mixing situations characterized by complex rheological behavior and viscosity evolution to name a few. A new coaxial mixer has been introduced by SHI Mechanical & Equipment, known as Superblend, which combines a Maxblend® impeller and a helical ribbon. The aim of this work was to experimentally investigate the performance of the Superblend, focusing on rotating modes (co- and counter-rotating) in the laminar, transition and turbulent flow regimes, and also to compare its performance with previously studied dual shaft coaxial mixers. In order to achieve the objective, a fully instrumented Superblend (40 L) was used, consisting of a Maxblend® impeller and a helical ribbon mounted on two independently actuated coaxial shafts running respectively at high and low speed and operating either in co-rotating or counter-rotating mode. Power consumption and mixing time experiments using the discoloration technique were conducted using aqueous solutions of corn syrup and water as Newtonian fluid. Based on mixing efficiency criteria, this study confirmed that the co-rotating mode is consistently yielding the best results in the laminar and early transition regimes. Recent power correlations introduced by our group for coaxial mixers were tested for their applicability and they were proven to be very applicable for the Superblend. This work demonstrates the efficiency of the Superblend and its superiority when compared to the dual shaft coaxial mixers based on different mixing efficiency criteria.

Keywords: Superblend, Maxblend, Helical Ribbon, Newtonian fluid, Laminar, Turbulent, Power consumption, Mixing time.

248c Flow and Shear In Agitated Vessels

Richard K. Grenville[1], James N. Tilton[1], Thomas A. Simpson[1], David A. R. Brown[2], Gustavo A. Padron[3], and Arthur W. Etchells III[4]. (1) DuPont Engineering Technology, DuPont Building D5054, 1007 Market St., Wilmington, DE 19898, (2) BHR Group, The Fluid Engineering Centre, Cranfield, MK43 0AJ, United Kingdom, (3) Fluid Engineering Centre, BHR Group Limited, Cranfield, Bedfordshire MK43 0AJ, United Kingdom, (4) AWE3 Enterprises, Philadelphia, PA 19106

Impellers are often described in qualitative terms using expressions such as "gentle", "low shear", high efficiency", "high shear", etc. Axial flow impellers, such as hydrofoils, are often considered gentle and low shear while radial flow impellers, such as disc turbines, are considered high shear. There is a class of bladeless impellers which are essentially discs or circular saw blades which operate at high rotational speeds and are used in "very high shear" applications.

Using the power and flow numbers, the energy input per unit mass of fluid E can be obtained by dividing power by the product of density and volumetric flow rate. Plotting E versus impeller to vessel diameter ratios shows that the impellers qualitatively said to be "high shear" have a higher value of E at the same power input per unit mass. Scaling-up at constant power per mass increases E for a given impeller. High shear impellers, having large E values also have greater turbulence intensity (ratio of fluctuating to mean velocity).

When droplet sizes generated by these impellers at the same mean power input per mass are considered, hydrofoils which are considered "low shear" produce a smaller droplet size than Rushton turbines which are "high shear". The reason for this is that the hydrofoils have a smaller turbulent length scale and, as a result, at a given mean power input per mass the local turbulence in the impeller region is more intense producing smaller droplets. But, the smaller length scale is a result of the shallow blade angle. This in turn results in a smaller power number which means that the hydrofoil must operate at a higher speed. At the same power per mass, the impeller with a higher tip speed produces a smaller droplet and vice versa (at the same tip speed, the impeller with a higher power input produces a smaller droplet).

Gross measures of the impeller characteristics, the Power and Flow numbers, can be used to quantify the flow and shear characteristics of impellers in agitated vessels but the fine details of the turbulence generated must also be considered in order to fully understand their performance.

2008 AIChE Annual Meeting
Philadelphia, PA
November 16 – 21, 2008

Momentous Flow Technology – the other way of mixing.

Todd Hutchinson[1], Binxin Wu[1] and Wojtek Wyczalkowski[1]
[1]Philadelphia Mixing Solutions, Ltd., 1221 E. Main Street, Palmyra, PA 17078, USA

ABSTRACT

The continuous rotating flow, referred to as Momentous Flow Technology, may find application in alumina precipitators and anaerobic digesters. This technology allows use of relatively inexpensive mixers for large tall form vessels above 6 m in diameter and with Z/T ~=3.0. Circular flow in a vessel is generated by an axial impeller located in the upper part of the vessel with no baffles. As a result of fast rotation, a tornado like vortex is generated in the vessel center. Lab experimentation in a 0.91 m diameter vessel confirmed that this vortex has the capacity of lifting and uniformly distributing alumina particles up to 63 microns in diameter. The same principle can be applied for economical mixing in anaerobic digesters. The design of full scale units has to be supported by CFD. The 0.29 m diameter model vessel was built to conduct PIV measurements and validate CFD calculations.

INTRODUCTION

The unsteady rotating flow field in a cylindrical vessel called "whirlpool" is widely used in the food industry as a method of separating particles out of suspension (cup of tea method). This phenomenon can be alternatively used as a method for suspension of particles that have density much greater than fluid. Experimental studies of unsteady flow field were performed by Greenspan (1969), Denk (1991) et al and Jakubowski (2007) et al.

The scope of this work was to investigate the possibility of application of the rotating flow for solids suspension in industrial applications and in particular for alumina precipitators and anaerobic sludge digesters. In both applications vessels were 6 m to 20 m in diameter with Z/T~=3.0. The full scale design needs to be supported by CFD simulation; therefore it is important to validate simulation results with measurements. This paper describes the experimental approach to this investigation.

The Momentous (rotating) Flow is induced by a rotating axial impeller located in the upper part of the vessel. Once the entire body of liquid is put in spin, the tornado like vortex is formed in the vessel center. This vortex has the capability of lifting solids particles up to the impeller level and above. After passing the impeller the particles are moved to vessel walls and slowly settle to the conical bottom where they are lifted up again.

In addition to suspension of solids the entire body of fluid is in fast rotational motion. There are no stagnant zones and no possibility of scaling at the walls. When Momentous Flow is applied to anaerobic digester, centrifugal forces push methane bubbles to the center of rotation where they quickly coalesce and escape from the liquid.

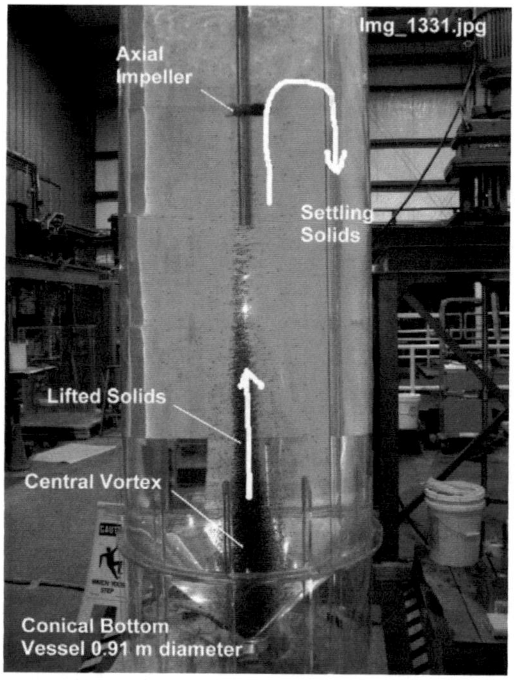

Fig 1. Experimental setup for solids lifting capacity in 0.91 m diameter vessel. Rotating flow in a vessel with no baffles is induced by a rotating impeller located in the upper part of the vessel. For solids suspension application, the bottom should be conical with apex angle no greater than 90°.

MODEL/EXPERIMENTAL

Experimental and CFD models were setup to provide the following information:

- Solids lifting capacity and energy usage. We have used alumina particles with three size ranges: 37-44 micrometers, 64-73 micrometers and 125-149 micrometers. Experiment has used 0.91 m diameter vessel with 2.75 m liquid level and conical bottom.
- CFD modeling of lab setup and full scale installation design. Reynolds Stress Model was used.

- Verification of CFD using lab setup and PIV. The 0.29 m diameter vessel was used for measurement of three components of velocity. Later the data was used for validation of CFD model.
- Determination of scale-up criteria using 0.29 m and 0.91 m vessels.
- Flow behavior in non-Newtonian fluids. Velocity measurements were taken for CMC solution in order to simulate anaerobic digester with 5% solids.

RESULTS

Solids lifting capacity was determined by visual observation. It was concluded during the experiment that 37-44 micrometers fraction was completely suspended off bottom. The 64-73 microns fraction was only partially suspended while the majority of solids were spinning in the conical bottom. The largest fraction of 125-149 micrometers was not suspended at all; however the solids were spinning around the bottom.

CFD Reynolds Stress model does well reproduce experimental findings. The axial motion of the vessel cone is well reproduced.

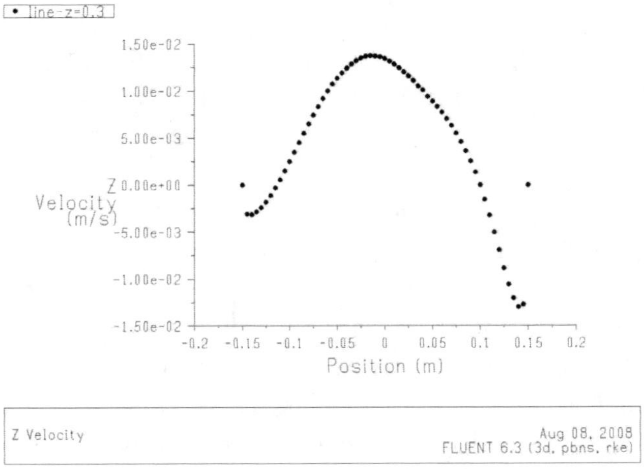

Fig. 2. Vertical component of velocity in rotating flow extracted from CFD at 0.3 m above bottom elevation in water

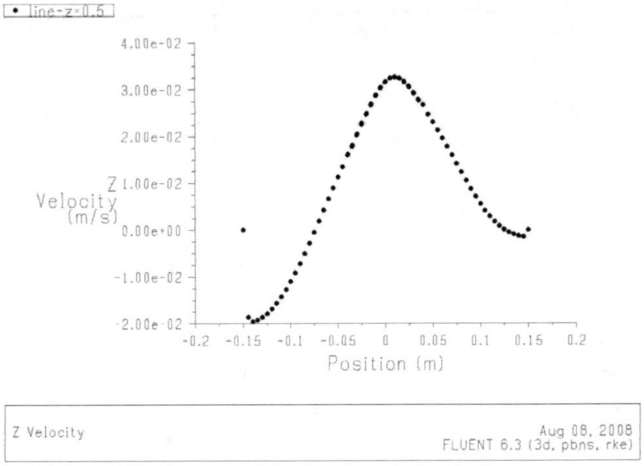

Fig. 3. Vertical component of velocity in rotating flow extracted from CFD at 0.5 m above bottom elevation in water

CONCLUSION

Based on lab test results the Momentous Flow Technology can be successfully applied for solids suspension in tall form vessels. Up pumping hydrofoil impeller has generated flow field that is capable of lifting alumina particles up to 63 microns in diameter. Larger particles can not be lifted using this technology regardless of mixer power.

Energy usage for solids suspension varies upon solids sizes and is in the range of 0.1 – 0.01 kW/m3

CFD Reynolds Stress Model can be used for flow study and illustration of process.

REFERENCES

Denk, V. and Durhold, A. (1991), "Experimental Investigation of the unsteady rotating flow field in a cylindrical vessel", Experiments in Fluids 12, 97-105

Jakubowski, M. and Diakun, J. (2007) "Simulation investigations of the effect of whirlpool dimensional ratios on the state of secondary whirls", Journal of Food Engineering 83, 1, November 106-110

Greenspan, H. P. (1969) "The Theory of Rotating Fluids" Cambridge University Press

Laser optical Measurements and numerical Investigations of Macro- and Micromixing in stirred Vessels

M. Faes, B. Glasmacher, Institute of Mutliphase Processes,
Leibniz Universitaet Hannover, Germany

Introduction

Today, increasing energy costs require effective processes, especially in chemical and pharmaceutical industries. Particularly, energy-intensive mixing processes offer high potential for optimization. Despite great efforts of research mixing of liquid-liquid components is not completely predictable. To construct mixing systems, empiricism is used rather than basic knowledge. Local mixing quality is a crucial parameter influencing the economics of such processes. Because of this interaction the interest of investigations concentrates on the measure of local mixing quality in regard to the macroscopic energy inserted mostly by stirrers. In case of a superimposed chemical reaction, complete mixing on molecular scale is required. The prediction of the local composition on molecular scale of a multicomponent system is not possible for complex geometries in mixing processes. A review of available models for mixing processes is given by Baldyga and Bourne (1999). The interaction between stretching, diffusion and reaction at small scales is described by Ottino (1994) in a lamellar model. Muzzio et al. (2003, 2002) extended a one-dimensional model to the complex flow in stirred tanks. Anderson et al. (2002) presented an extended mapping technique for chaotic flows in a journal bearing flow. Gollub et al. (2003) experimentally determined the mixing for a two-dimensional time-periodic flow exhibiting chaotic mixing. Despite the progress in predicting reactive mixing there is still a need for experiments, visualizing the local distribution of inert and reactive tracers with high spatial and temporal resolution. Laser induced fluorescence has proven to be a suitable measurement technique by Kling and Mewes (2003). The visualization of passive scalars indicates the convective mixing process (Villermaux et al. (1996), Distelhoff and Marquis (2000), Guillard et al. (2000)). In viscous mixing applications stretching and folding occur simultaneously at different rates in each portion of the flow, creating complex, layered structures. A seminal paper concerning the concept of chaotic advection has been published by Aref (1984). Reactive tracers which are pH-sensitive are used for measuring the micromixing indirectly by Bellerose and Rogers (1994) and Hong et al. (2002). In T-shaped micro-mixers the transport phenomena on different scales are investigated by means of pH-sensitive tracers using scanning methods by Hoffman et al. (2006). Van Vliet et al. (2000) used also a laser scanning method for investigating three dimensional concentration fields.

In order to visualize local compositions and dissipations the described research project comprises two laser optical measurement techniques with high spatial and temporal resolution. These are published by Faes and Glasmacher (2007, 2008). An investigation of local mixing quality, local energy dissipation and their correlation is performed. Therefore, the 4D-LIF technique is developed for analyzing the concentration field in dependence of time and a PIV system for analyzing the velocity and dissipation field. For the prediction of mixing quality in laminar flow, numerical calculations are performed for comparable experimental parameters.

Experimental Set-Up

The laser optical experiments are performed in a cylindrical glass vessel, which has an inner diameter of d_a = 90 mm. For the investigations a multi-stage pitched blade stirrer is used. A rectangular viewbox filled with pure glycerol and positioned round the mixing system allows an isooptical study and minimizes optical distortions. The working fluid is a mixture consisting of pure glycerol and a solution of calcium chloride. It is possible to measure velocity and concentration fields throughout the whole mixing system. The viscosity of the glycerol mixture is 0.93 Pas measured with Rheometrics RFS II at room temperature. The experimental set-up, consisting of the two optical measurement techniques, is schematically depicted in figure 1. The particle image velocimetry, a LaVision system, is used to measure local velocity vectors. The laser is a double pulsed NewWave® Nd:YAG laser (λ = 532 nm) with a frequency of 10 Hz. The illuminated laser plane can be adjusted freely by an automated linear positioning system. The illuminated horizontal laser plane is projected to the CCD-camera by means of a mirror under 45°, which is placed below the viewbox. The camera and laser control, data acquisition and processing is performed with a software package DaVis® from LaVision. For measuring the velocity field (PIV), the particle paths are visualized by seeding the working fluid with hollow glass spheres of mean diameter of 10 μm. The reflected signals of the glass spheres are captured temporally with a CCD-camera, Imager Intense 1376 x 1040 pixel. For the image processing a cross-correlation algorithm with an interrogation window of 32 x 32 pixel with 50% overlap is selected. In each case 200-500 double images are captured. The second non-intrusive optical laser measurement technique is the 4D two-color laser induced fluorescence technique. It is used to measure the concentration fields of two fluorescent dyes simultaneously as described by Kling (2004). The laser is a NewWave® Nd:YAG (λ = 495 nm), Tempest 30 with GWU OPO® VisIr. The CCD-camera, Imager 3 (640 x 480 pixel) and Image Intensifier delivered by LaVision, is equipped with a so-called Double-Image Optics (DIO), which permits to capture the same display window twice at the same time. The optical DIO consists of two apertures, which are equipped with a set of adjustable and fixed mirrors, and two optical filters (BP523/10 and RG645) for separating the fluorescent lights of the dyes. For the multi-stage pitched blade stirrer a volume of 2 ml is injected within 10 s. The illuminated laser planes have a thickness of 500 μm. For receiving three-dimensional images in dependence of time (4D) the cameras, the mixing vessel and the light sheet optics are mounted onto a linear positioning system. In order to measure the concentration field, the stirrer is stopped. The three-dimensional images are reconstructed out of 170 two-dimensional plane images, captured in a distance of 500 μm, using Imaris®-Software from Bitplane. Each two-dimensional plane image is an averaged plane image out of 16 single plane images captured in one second. The scanning procedure from the bottom of the mixing vessel to the free surface of the working fluid takes 10 minutes including post processing. The laser measurement technique of PIV and 4D-LIF are used sequentially.

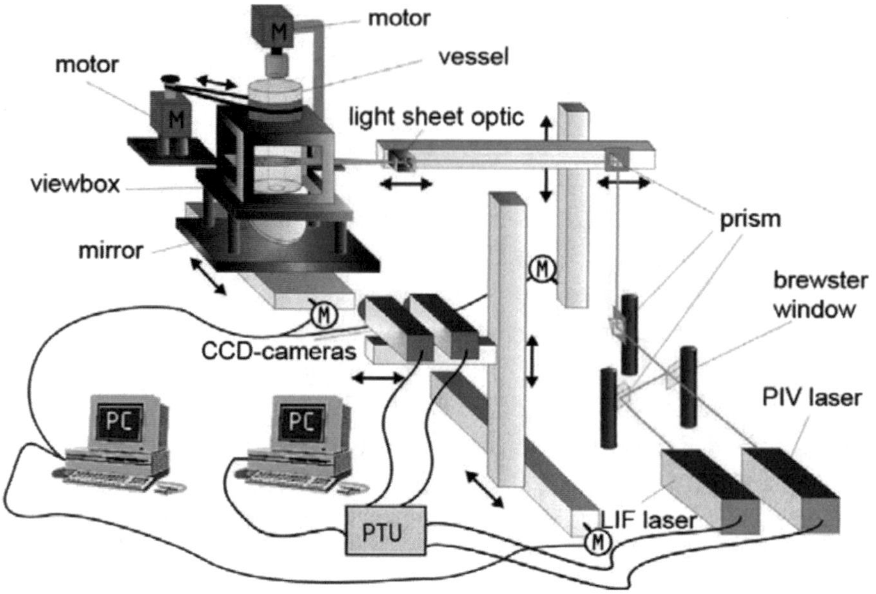

Figure 1: Measurement techniques consisting of 4D-LIF and PIV

Macro- and Micromixing

The convection, namely the macromixing, is visualized by an inert fluorescent dye (carboxy-SNARF). The transport on molecular scale is visualized by a fluorescent dye (fluo-4) reacting with calcium ions dissolved in the working fluid. The fluorescent emission characteristic of this dye changes due to a chemical reaction with calcium ions. The progress of the chemical reaction shows indirectly the micromixing, because mixing on molecular scale is required for the progress of chemical reactions. After a mixture of both dyes is injected into the mixing system, the concentration fields of the inert and reactive dyes are measured. The fluorescence intensity of the dyes are calculated by Lambert-Beer's law. The quantum yield Φ describes the effectiveness of the fluorescent emission I_F:

$$I_F = \Phi I = \Phi I_0 \exp(-\varepsilon s c) \qquad (1)$$

I_0 is the intensity of the exciting light, ε is the molar extinction coefficient and s is the length of the measurement volume. A simplification by series expansion can be done for small concentrations so that I_F depends linearly on the concentration c of the dye:

$$I_F = \Phi I_0 K \varepsilon s c = mc \qquad (2)$$

K is a parameter depending on the measurement system, considering for example the viewing angle of the detector. For constant parameters, m is a calibration factor. This allows a correlation of local dye concentrations and measured intensities. More details, especially the calibration of the measurement system and the correction of the shot-to-shot noise of the laser, are described in Kling and Mewes (2004).

The local degree of deviation is a quantitative measure for the quality of micromixing. It is defined as

$$\Delta(\vec{x},t) = 1 - \frac{c_{1,react}(\vec{x},t)}{c_1(\vec{x},t)} \tag{3}$$

and can be calculated by comparing the local concentration of the reaction product $c_{1,react}$ with the virtual concentration c_1 of the reacting dye locally appearing without any reaction. Due to the injection as a mixture and transportation in the same manner the local concentration of the reactive dye c_1 without chemical reaction can be calculated by the knowledge of concentration c_2 of the inert dye and the initial concentration ratio to

$$c_1(\vec{x},t) = c_2(\vec{x},t)\frac{c_{1,0}}{c_{2,0}}. \tag{4}$$

The local degree of deviation is the fraction of the reacting dye which has not reacted yet. For a value of the local degree of deviation $\Delta = 1$ the mixture is completely segregated on macroscopic scale. During the mixing process a homogenous fluid on microscopic scale is achieved and the local degree of deviation decreases to $\Delta = 0$ for a completely micromixed fluid.

Numerical Methods

For mixing processes it is desirable to have local information about concentration and velocity fields to optimize the mixing geometry or the position of the feeding point. For a theoretical prediction of local mixing quality it is necessary to choose the correct physical models and boundary conditions. In this work, the data of the numerical calculations are validated by the experimental investigations. The numerical calculations are performed with a commercial CFD code by ANSYS. The geometrical models are equivalent to the glass mixing vessel equipped with the full-glass stirrer in the experimental set-up. The meshes of the stirrer configuration consist of 6.0 million finite volumes for the multi-stage pitched blade stirrer. The validated results serve for the reduction of experiments and their high costs.

Results and Discussion

Multi-stage pitched blade stirrer

The speed of each stirrer is adjusted due to the chosen Reynolds numbers for laminar conditions. The mixing vessel is filled with 500 ml of the working fluid. Due to a complete field of viewing of the uniform secondary flow field, the distance from the bottom of the mixing vessel to the centre of the lower blade of multi-stage pitched blade stirrer is set to 25 mm as depicted in figure 2. The distance between the centers of the two stages is 30 mm. The multi-stage pitched blade stirrer induces a complex three-dimensional flow field with three torus-shaped eddies. The structure of the eddies depends of the rotational speed of the stirrer. Here, three different injection positions are analyzed, which are depicted in figure 2 as A,B, and C. The progress of macro- and micromixing is depicted in figure 3 for injection position B.

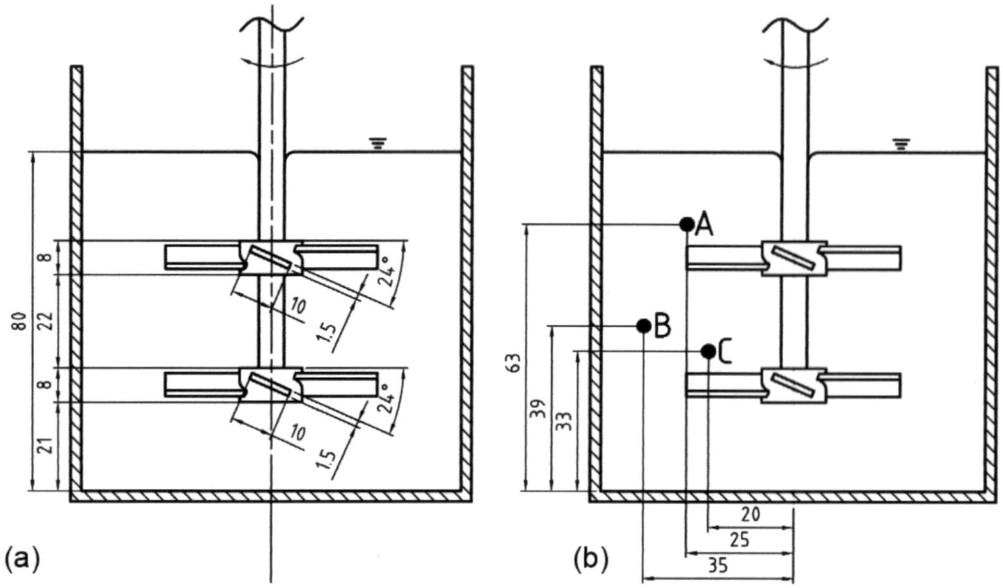

Figure 2: Schematic of the multi-stage pitched blade stirrer (a), and of the three injection positions A, B, and C (b)

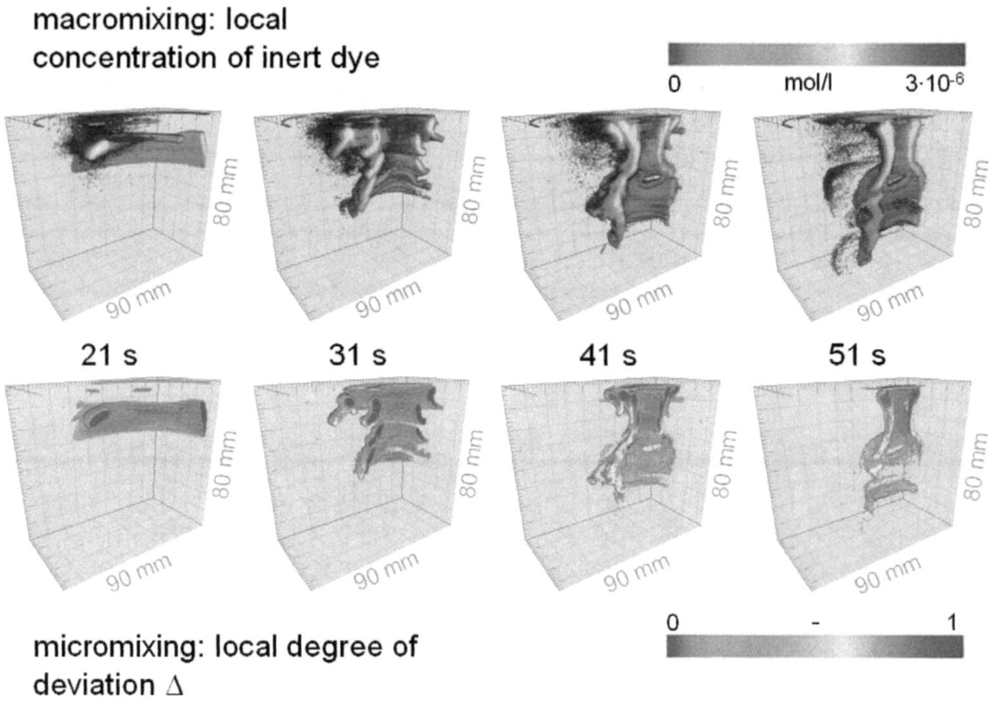

Figure 3: Progress of macro- and micromixing for a multi-stage pitched blade stirrer, Re = 3, injection position B

After the injection of the dye mixture, lamellar structures are formed and transported upwards before they reach the two stages of the downward-downward pumping stirrer. The striation thickness of the lamellas decreases due to the convective transport of the volume. The progress of the micromixing proceeds simultaneously. The local degree of deviation decreases from the starting value of 1, which is indicated by red color. The dye volume transported into high regions of dissipation near the blades of the stirrer is micromixed well. The local degree of deviation decreases. A quantitative analysis is given during the oral presentation.

Conclusion

Macro- and micromixing in laminar flow regime is analyzed by experimental and numerical methods. The quality of the mixing process is described by the local degree of deviation, calculated by the local concentrations of an inert and a reactive dye. The progress of mixing is observed as a function of time for different injection positions and Reynolds numbers. In a complex flow field, induced by a multi-stage pitched blade stirrer, local concentrations are measured and visualized four-dimensionally. Mixing on molecular scale depends mainly on the flow path through the mixing vessel and the inserted local energy dissipation. Micromixing starts in the boundary layers of the lamellas, indicated by a decreasing local degree of deviation. The developed numerical model provides the prediction of macro- and micromixing. The numerical calculations show good accordance with the experimental investigations. The developed methods and investigations in this project facilitate the understanding of transport phenomena in liquid-liquid mixing systems for laminar conditions.

Acknowledgement

The research project is supported by the German Research Foundation (DFG) within the priority program SPP 1141 "Analysis, Modeling and Calculation of Mixing Processes with and without Chemical Reaction". We gratefully acknowledge the financial grant and the scientific support by the chair D. Mewes and the precursor project of K. Kling and D. Mewes.

References

Anderson, P. D. et al. (2002), "Material streching in laminar mixing flows: extended mapping technique applied to the journal bearing flow", International. Journal of Numerical Methods in Fluids, 40, 189-196

Aref, H. (1984), "Stirring by chaotic advection", Journal of Fluid Mechanics, 143, 1-21

Baldyga, J., Bourne, J.R. (1999), "Turbulent mixing and chemical reaction", John Wiley & Sons Ltd

Bellerose, J.A., Rogers, C.B. (1994), "Measuring mixing and local pH through Laser Induced Fluorescence", Laser Anemometry, 191, 217-220

Bird, R.B. et al. (2002), "Transport Phenomena", John Wiley and Sons Inc., New York

Distelhoff, M.F.W., Marquis, A. J. (2000), "Scalar mixing in the vicinity of two disk turbines and two pitched blade impellers", Chemical Engineering Science, 55, 1905-1920

Faes, M., Glasmacher, B. (2007), "Analyse des Makro- und Mikromischens mit Hilfe von laserdiagnostischen Verfahren", Chemie Ingenieur Technik 79, 2007, No.7, 1059-1065

Faes, M., Glasmacher, B. (2007), "Visualisieren der Transportvorgänge beim Makro- und Mikromischen mit Hilfe der laserinduzierten Fluoreszenz", Chemie Ingenieur Technik 79, 2007, No.9, 1468

Faes, M., Glasmacher, B. (2008), "Measurements of micro- and macromixing in liquid mixtures of reacting components using two-colour laser induced fluorescence", Chemical Engineering Science, 63, 4649-4655
Faes, M. et. al. (2008), "Mehrskalige Analyse von Transportprozessen mit Hilfe von laseroptischen Methoden und CFD", Chemie Ingenieur Technik 80, 2008, No.9, 1287
Gollub, J. P. et al. (2003), "Mixing rates and symmetry breaking in two-dimensional chaotic flow", Physics of Fluids, 15, 9, 2560-2566
Guillard, F. et al. (2000), "New image analysis methods for the study of mixing patterns in stirred tanks", Canadian Journal of Chemical Engineering, 78, 273-285
Haugland, R.P (Edt.) (2002), Molecular Probes- Handbook of Fluorescent Probes and Research Products, 9th Edition
Hoffmann, M. et al. (2006), "Experimental investigation of liquid-liquid mixing in T-shaped micro-mixers μ-LIF and μ-PIV", Chemical Engineering Science, 61, 2968-2676
Hong, S. D. et al. (2002) Proceedings of the 10th International Symposium on Flow Visualization, August 26-29, 2002 in Kyoto, Japan, paper F0275
Kling, K., Mewes, D. (2003), "Quantitative Measurements of Micro- and Macromixing in Stirred Vessels using Planar Laser Induced Fluorescence", Journal of Visualization, 6, 165-173
Kling, K., Mewes, D. (2004), "Two-colour laser induced fluorescence for the quantification of micro- and macromixing in stirred vessels", Chemical Engineering Science, 59, 1523-1528
Kling, K. (2004), "Visualisieren des Mikro- und Makromischens mit Hilfe zweier fluoreszierender und chemisch reagierender Farbstoffe", Dissertation, Universität Hannover
Muzzio, F.J. et al. (2002), "Mechanisms of mixing and creation of structure in laminar stirred tanks", AIChE Journal, 48, 2135-2148
Muzzio, F.J., et al. (2003), "Effect of hydrodynamics on reactive mixing in laminar flows", AIChE Journal, 49, 168-179
Naraghi, M. (1997), "T-jump study of calcium binding kinetics of calcium chelators", Cell Calcium, 22, 255-268
Ottino, J.M. (1994), "Mixing and chemical reactions - a tutorial", Chemical Engineering Science, 49, 4005-4027
Villermaux, J. et al. (1996), "Planar laser induced fluorescence technique for measurements of concentration fields in continuous stirred tank reactors", Experiments in Fluids, 22, 95-102
Van Vliet et al. (2000), "Four-dimensional Laser Induced Fluorescence measurement of micro mixing in a tubular reactor", Proceedings of the 10th European Conference on Mixing, 45-52

248f Effective Shear Stress Model as a Unified Concept for the Mixing of Non-Newtonian Fluids In the Laminar, Intermediate and Turbulent Region

Karl-Heinz Wassmer, Polymer Research, Polymer Technology, BASF SE, GKE/S - B1, Ludwigshafen, Germany

Stirring is one of the most important tasks in chemical, pulp, polymer and food industries. It determines process parameters like power consumption, mixing time, heat transfer etc. Usually dimensionless numbers like Newton and Reynold numbers are used to characterize the systems, where the latter one is strongly dependent on viscosity. Stirring usually creates a distribution of shear rates in a tank. For Newtonian fluids, where the viscosity does not depend on the shear rate, the calculation of the Reynolds number is straightforward when using a constant viscosity.

However, in the case of non-Newtonian fluid behavior the shear field within a stirred tank corresponds to a viscosity field. In the laminar flow region the concept of effective viscosity by Metzner and Otto is well established. In the transition region between laminar and turbulent flow, the existing concepts use three or even more empirical parameters to determine the specific power input. The effective shear stress concept is a general but straightforward approach to calculate the power input for shear thinning fluids over the whole flow region by introducing only one shear coefficient as an empirical parameter. The Metzner-Otto relation is obtained as a limiting case for the laminar region. The shear coefficient of the new approach is related to the Metzner-Otto constant and peculiarities of the Metzner-Otto concept can be explained. The concept is validated for eight different stirrer systems. Mixing time, maximum shear rate and heat transfer can also be calculated using this approach. The presented new concept is also applicable for other apparatuses, e.g. static mixers and pipes.

301a New Perspectives for Reactive Mixing In Multiphase Flows

Stefan Radl and Johannes G. Khinast. Institute for Process Engineering, Graz University of Technology, Inffeldgasse 21a, Graz, 8010, Austria

Fast chemical reactions in sparged stirred tanks are of great importance in the chemical industry. If multiple reaction steps occur within one apparatus, a detailed understanding of mixing is crucial to precisely control the reaction network. Clearly, a quantitatively correct model of the interaction between reactions, mass transfer and liquid phase mixing is necessary to attain this goal.

Although there has been significant experimental and theoretical work in this field, fundamental challenges still remain. For example, fully resolved 3D simulations of reacting bubble swarms with complex reactions are still extremely demanding with respect to computational resources. Hence, our previous work[1-3] was based on 2D direct numerical simulations of bubbly flows and three-dimensional effects were ignored completely.

In this work it is our intension to critically prove our previous assumption by using both theoretical and numerical arguments. Therefore we use different simulation approaches that allows us to simulate bubbly flows including mixing and reactions on different scales in full 3D. Our work includes direct numerical simulation of bubble wake phenomena and large eddy simulations to study bubble plume effects. The results of these simulations enable us for the first time to characterize the impact of mixing on the reaction network in different regions of a bubble plume. Finally, we propose new perspectives on how to model the complex dynamics in bubble swarms when fast reactions are involved.

(1) Radl S, Tryggvason G, Khinast JG. Flow and Mass Transfer of Fully Resolved Bubbles in Non-Newtonian Fluids. AIChE Journal. 2007;53:1861-1878.

(2) Radl S, Khinast JG. Prediction of mass transfer coefficients in non-Newtonian fermentation media using first-principles methods. Biotechnology and Bioengineering. 2007;97:1329-1334.

(3) Radl S, Koynov A, Tryggvason G, Khinast JG. DNS-based Prediction of the Selectivity of Fast Multiphase Reactions: Hydrogenation of Nitroarenes. Chem Eng Sci. 2008; in press.

Effects of Oxygen and Air Mixing on Void Fractions in a Large Scale System

Robert A. Leishear, Savannah River National Laboratory, Aiken, SC

Michael L. Restivo, Savannah River National Laboratory, Aiken, SC

Hector N. Guerrero, Savannah River National Laboratory, Aiken, SC

David J. Sherwood, Hanford Waste Treatment & Immobilization Plant Project, Richland, Wa.

Abstract

Oxygen and air mixing with spargers was performed in a 30 foot tall by 30 inch diameter column, to investigate mass transfer as air sparged up through the column and removed saturated oxygen from solution. The mixing techniques required to support this research are the focus of this paper. The fluids tested included water, water with an anti-foam agent (AFA), and a high, solids content, Bingham plastic, nuclear waste simulant with AFA, referred to as AZ01 simulant, which is non-radioactive. Mixing of fluids in the column was performed using a recirculation system and an air sparger. The re-circulation system consisted of the column, a re-circulating pump, and associated piping. The air sparger was fabricated from a two inch diameter pipe concentrically installed in the column and open near the bottom of the column. The column contents were slowly re-circulated while fluids were mixed with the air sparger. Samples were rheologically tested to ensure effective mixing, as required. Once the fluids were adequately mixed, oxygen was homogeneously added through the re-circulation loop using a sintered metal oxygen sparger followed by a static mixer. Then the air sparger was re-actuated to remove oxygen from solution as air bubbled up through solution. To monitor mixing effectiveness several variables were monitored, which included flow rates, oxygen concentration, differential pressures along the column height, fluid levels, and void fractions, which are defined as the percent of dissolved gas divided by the total volume of gas and liquid. Research showed that mixing was uniform for water and water with AFA, but mixing for the AZ101 fluid was far more complex. Although mixing of AZ101 was uniform throughout most of the column, gas entrapment and settling of solids significantly affected test results. The detailed test results presented here provide some insight into the complexities of mixing and void fractions for different fluids and how the mixing process itself affects void fractions in Bingham plastic fluids, which have a measurable yield stress.

This document was prepared in conjunction with work accomplished under Contract No. DE-AC09-08SR22470 with the U.S. Department of Energy. This work was prepared under an agreement with and funded by the U.S. Government. Neither the U. S. Government or its employees, nor any of its contractors, subcontractors or their employees, makes any express or implied: 1. warranty or assumes any legal liability for the accuracy, completeness, or for the use or results of such use of any information, product, or process disclosed; or 2. representation that such use or results of such use would not infringe privately owned rights; or 3. endorsement or recommendation of any specifically identified commercial product, process, or service. Any views and opinions of authors expressed in this work do not necessarily state or reflect those of the United States Government, or its contractors, or subcontractors.

Introduction

This paper focuses on void fractions and mixing of fluids in an 8.4 meter tall by 0.76 meter diameter bubble column This paper is the last of three papers which describe mass transfer tests performed in this system, using various fluids[1,2,3]. Void fraction data was obtained for 24 different test conditions, at different fill levels in the bubble column. Superficial sparging velocities of 2, 5, and 10 mm/second were applied through the central sparger to each of the simulants at three different column fill levels, where the superficial velocity is defined as the average volumetric flow rate divided by the liquid surface area in the column. The primary fluid of concern was a Bingham plastic fluid, but mixing effectiveness was first demonstrated using Newtonian fluids, which were water and water with AFA added. Void fractions were one of the primary interests of this research, where void fractions are defined as the volume of gas divided by the total solution volume while bubbles rise in solution. This paper provides experimental void fraction data, and compares experimental results to theory. To do so, a system description is provided, followed by a discussion of mixing techniques, and a discussion of void fractions.

System Description

The primary mixing system consists of a bubble column, a centrifugal pump, and associated piping, supported by auxiliary components to fill the column and aid in mixing. A simplified schematic of the system is shown in Figure 1, a complete system is shown in Figure 2, and the installed bubble column is shown in Figure 3. Detailed component descriptions are provided as required throughout the paper, material properties of the tested fluids are summarized in Table 1. The fluids included process water, process water mixed with anti-foam agent which was cloudy white in color (Dow Corning, Q2-3183, antifoam agent), and AZ101 simulant which was reddish brown in color and is a mixture of chemicals used to simulate nuclear waste, which also included anti-foam agent.

Table 1: Fluid Properties

Fluid	Yield Stress, Pascals	Consistency / Viscosity, Centipoise
Water	0	1
Water with AFA	0	1
AZ101 with AFA	13	11.5
AZ101 with AFA	30	27.7

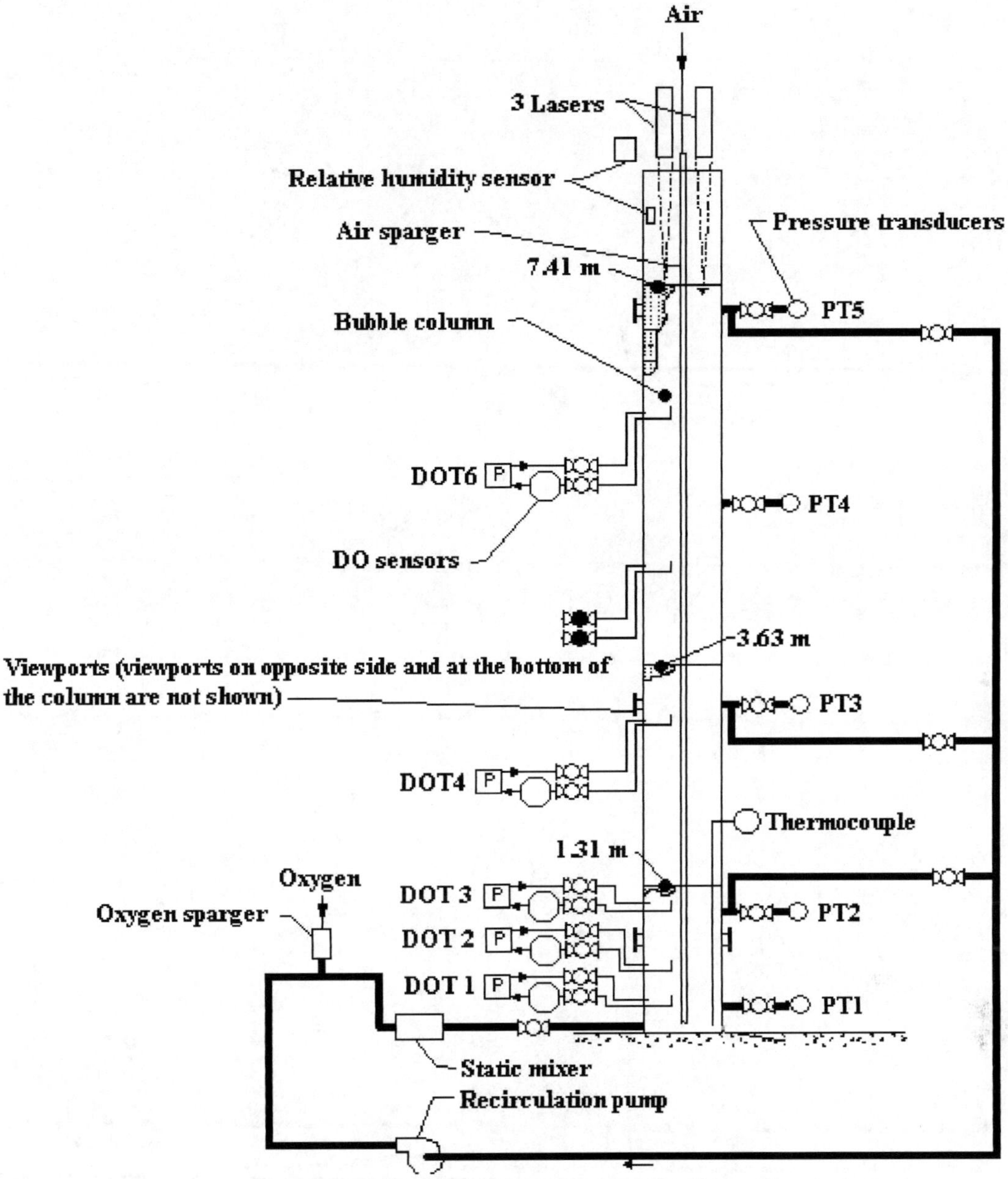

Figure 1: Simplified System Schematic[1]

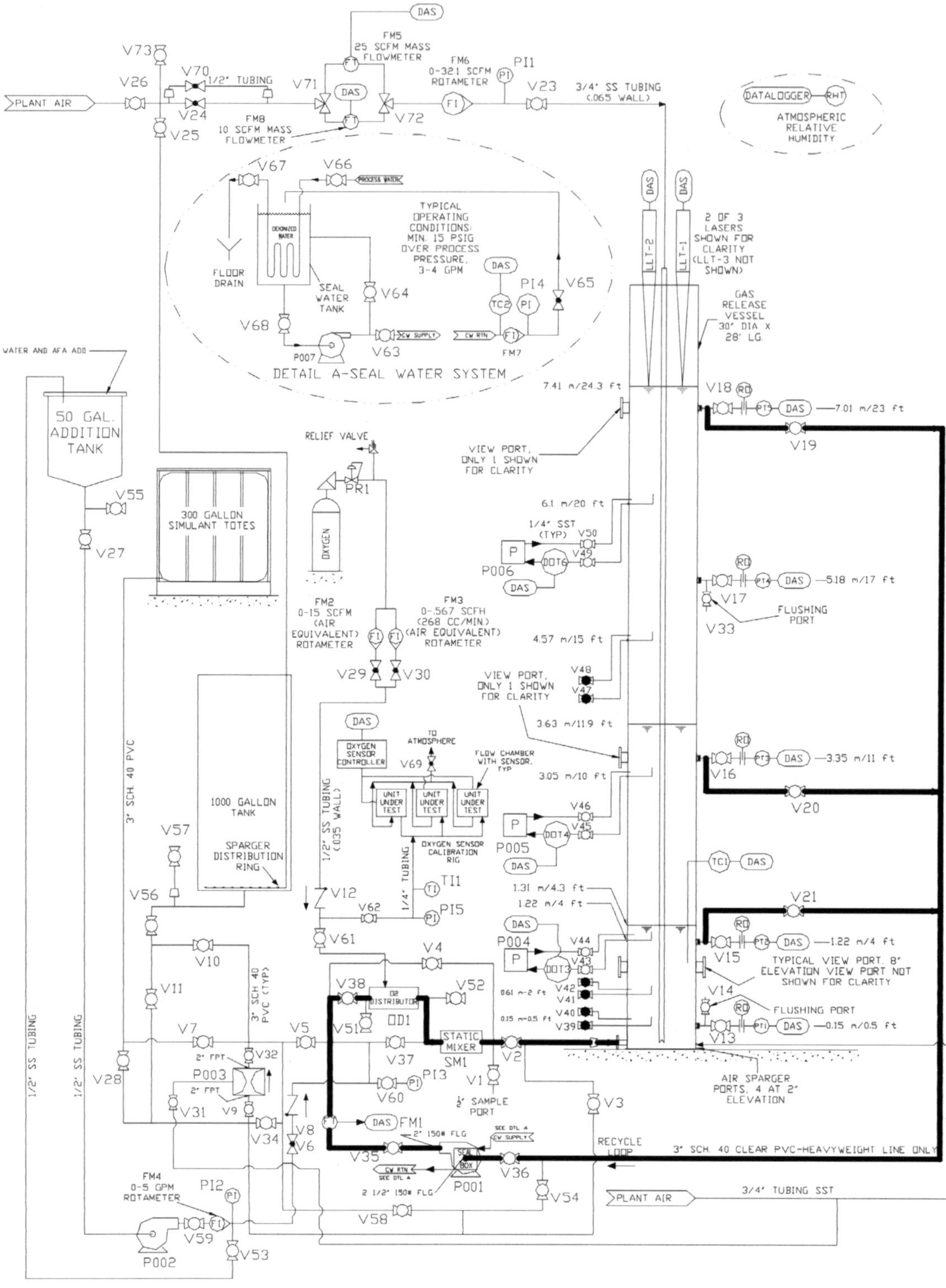

ELECTRONIC P&ID, REVISED 8/15/2007

Figure 2: System Schematic[3]

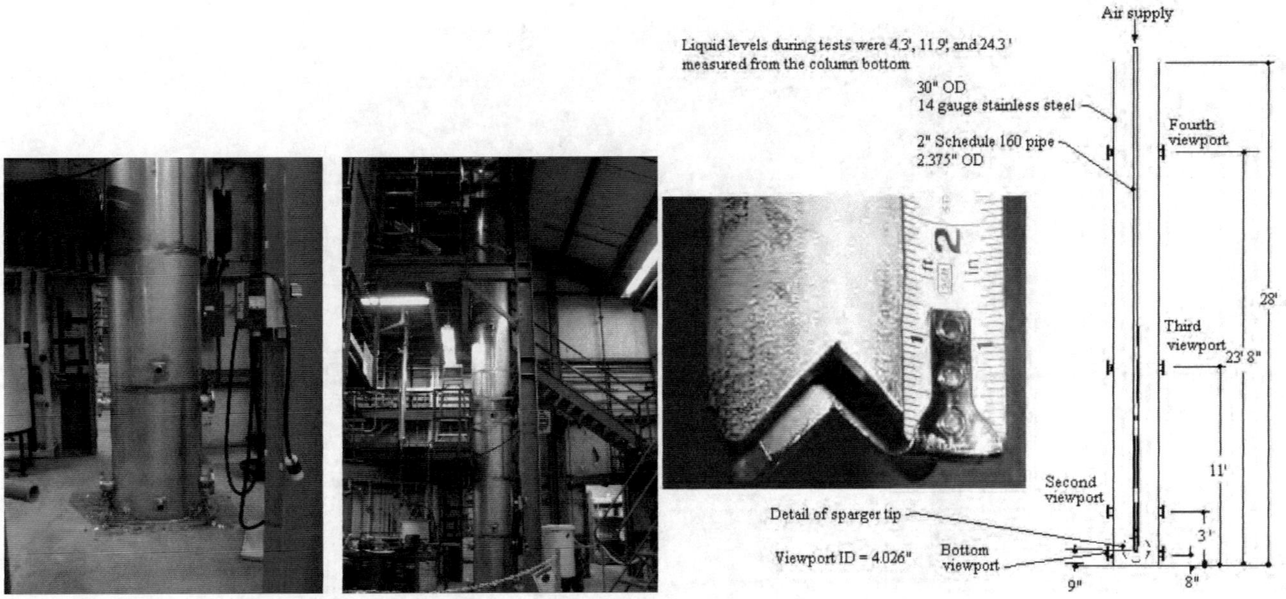

Figure 3: Bubble Column Design and Installation[1]

Mixing

There were four separate mixing processes required for these tests. One process added AFA to either water or AZ101 simulant as required. A second process thoroughly mixed the test fluids prior to testing. The third process saturated the test fluid with oxygen, and the fourth process removed oxygen from solution by sparging with air. Utilizing these processes, tests were performed at three different liquid levels in the bubble column, as shown in Figure 1 (1.31 meters, 3.63 meters, and 7.41 meters).

Adding AFA

To effectively mix AFA into solution, a common practice of simply mixing the contents for a total of four volume changes was employed. Referring to the left hand side of Figure1, the re-circulating pump, P002, was used to re-circulate the contents of the 50 gallon addition tank. Anti-foam agent (1.2 kg) was added to 27 gallons of water and re-circulated at 5 gpm to ensure that four volume changes occurred. Once the solution was well mixed in the addition tank, the AFA was added to the primary mixing system, which is shown in Figure 2 as the line shown as a heavy line. The re-circulating pump for this system operated at 100 gpm while the AFA was added at a uniform rate to the fully mixed test fluid, and the bubble column contents were then re-circulated for four volume turnovers after the AFA was added. When AFA was added to water, mixing effectiveness was easily verified. A transparent pipe is vertically attached to the side of the bubble column, and the column contents were mixed near the top of this column through this pipe just below the 7.41 meter column level, as shown in Figure 1. Visual inspection up the length of the transparent, PVC pipe verified uniform mixing, as shown in Figure 4. Once the AFA was mixed with water, the solution remained well mixed. Similarly, AFA was added to the opaque AZ101 solution, which is shown in Figure 4.

Figure 4: Bubble Column and Transparent External Piping

Initial Mixing of AZ101 Simulant

To ensure a homogeneous mixture before testing of the AZ101 simulant, effective mixing was required. To do so, the tank contents were re-circulated at 100 gpm for four volume turnovers. While the contents were re-circulating, the air sparger in the center of the column injected 20 scfm (10.46 scfm = 10 mm/second superficial velocity) of air and a circular sparging ring at the bottom the column (not shown) injected another 20 scfm of air. Effective mixing was verified by taking samples from the column at equal time increments, and then measuring the yield stress and consistency to ensure that properties were consistent between samples. Once an adequate mixing time was established, all further testing was preceded by mixing at the established time.

Mixing of Oxygen into Solution

Oxygen mixing was verified by measuring the dissolved oxygen concentration in solution, but mixing of oxygen in water helped verify the mixing process, since it was the only transparent fluid used in the tests. Oxygen was added to the water as it re-circulated at 100 gpm through the re-circulating pump and the column. Oxygen mixing was accomplished by using both a sintered metal filter which can create air bubbles down to 1 - 2 micron diameters and a helical static mixer which can create bubbles down to 20 – 30 microns. To ensure that air bubbles were as small as possible and that the bubbles were uniformly distributed, these two components were used in series, as shown in Figure 1. Oxygen was added at 3 scfm, and mixing of oxygen in the three inch diameter pipe is shown in Figures 3 – 5. Once the oxygen entered the column from the three inch pipe, bubbles almost immediately coalesced and formed larger bubbles of ¼ to ½ inch in diameter near the bottom of the column, as shown in Figure 7. The bubbles then rose to the top of the column in approximately 30 seconds and became a more uniform size of 1/8 to ¼ inch in diameter as they rose.

The 3 scfm oxygen flow rate was optimized based on measurements of dissolved oxygen (DO) concentrations in the bubble column. Dissolved oxygen sensors were installed at various elevations, as shown in Figure 1, but only three sensors were available and they were moved to different locations for different tests. To validate mixing, the dissolved oxygen concentration was monitored at three different column levels until the DO concentration reached equilibrium at each of those levels. Typical test results are shown in Figure 8.

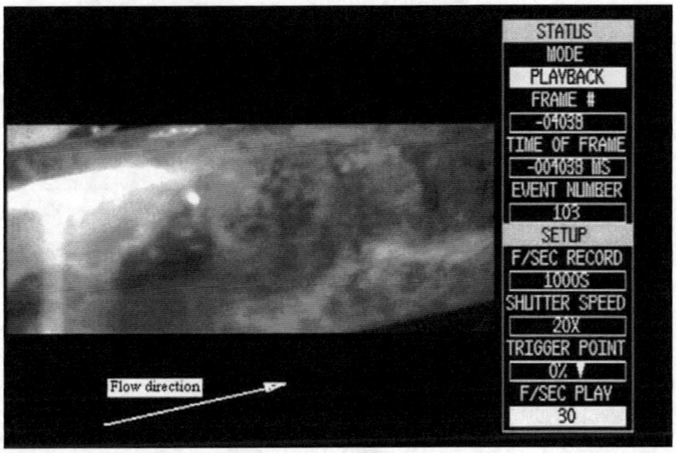

Figure 5: Bubbles Prior to the Static Mixer[1]

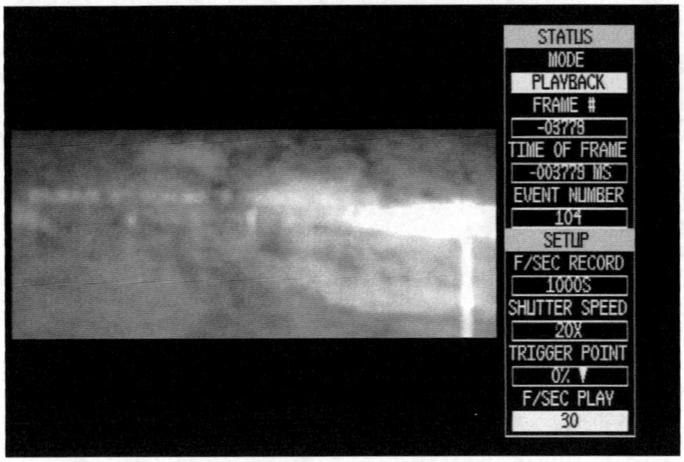

Figure 6: Uniform Distribution of Bubbles Leaving the Static Mixer[1]

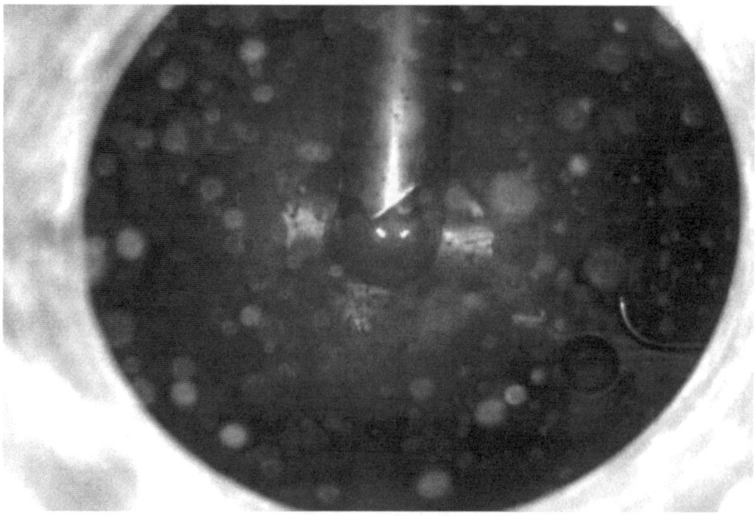

Figure 7: Bubbles Exiting the Mixing pipe Near the Column Bottom

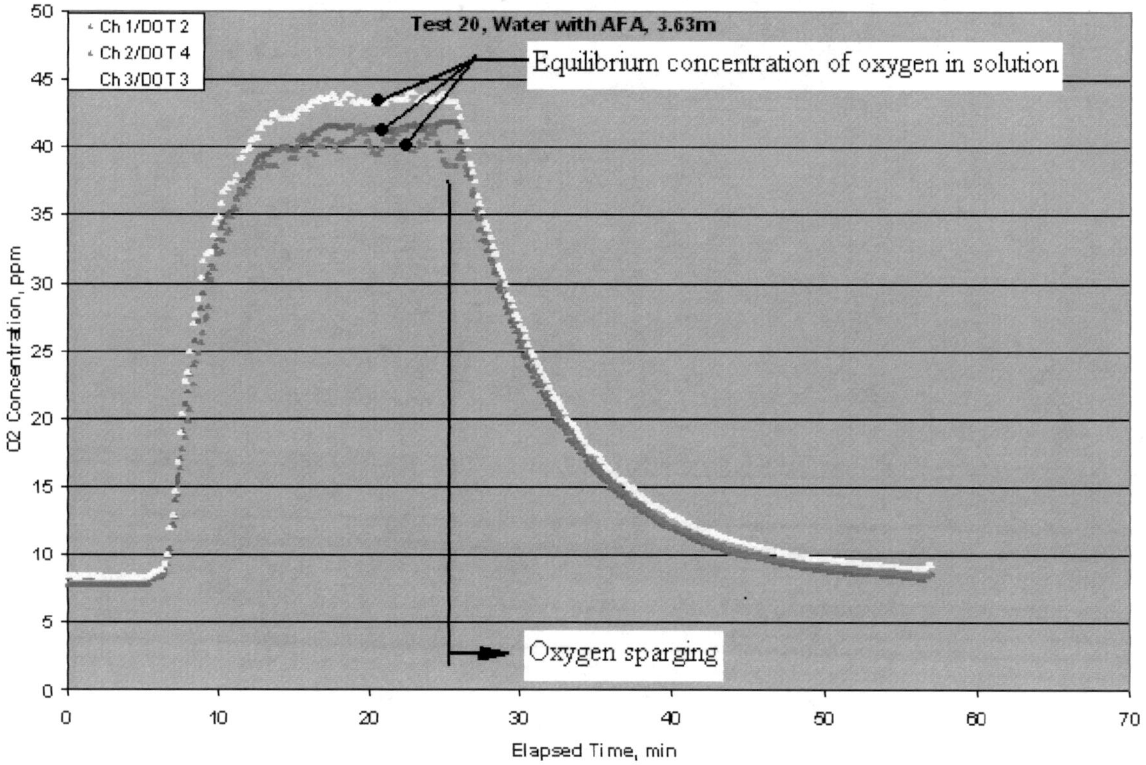

Figure 8: Typical Mass Transfer Test Results

Mixing of the Column Using the Air Sparger

To mix the column at three required superficial velocities, only the central sparger was used, and again DO concentrations were used to monitor mixing as shown in Fig. 8. For the water and water with AFA, mixing was consistent since the fluids were homogeneous. However, the AZ101 properties changed during testing due to fast settling solids in solution. Overnight, the simulant typically settled, and an inch of water separated from solution on the liquid surface in the column. During tests, it was obvious from the DO readings that the material settled during tests and affected the DO sensor readings. As the material settled, oxygen was trapped near the bottom of the tank, and bubbled out sporadically. Even though the material was thoroughly mixed at the beginning of the tests, this mixing problem persisted. The sparger was inadequate to mix the simulant near the column bottom, since the bubbles exiting the sparger formed a rising cone of bubbles. Mixing was uniform at higher elevations in the column since the mixing bubbles formed a uniform field of bubbles as they rose to the top of the column. To demonstrate this statement, typical photos of bubble formation in the column are shown in Figures 9 and 10 for water. Similar results were obtained for the AZ101 simulant, but photos were only available of the fluid surface since it was opaque, as shown in Figure 11. Even though the bubble sizes were markedly different, the DO concentrations proved that mixing was effective throughout most of the column length, except at the bottom.

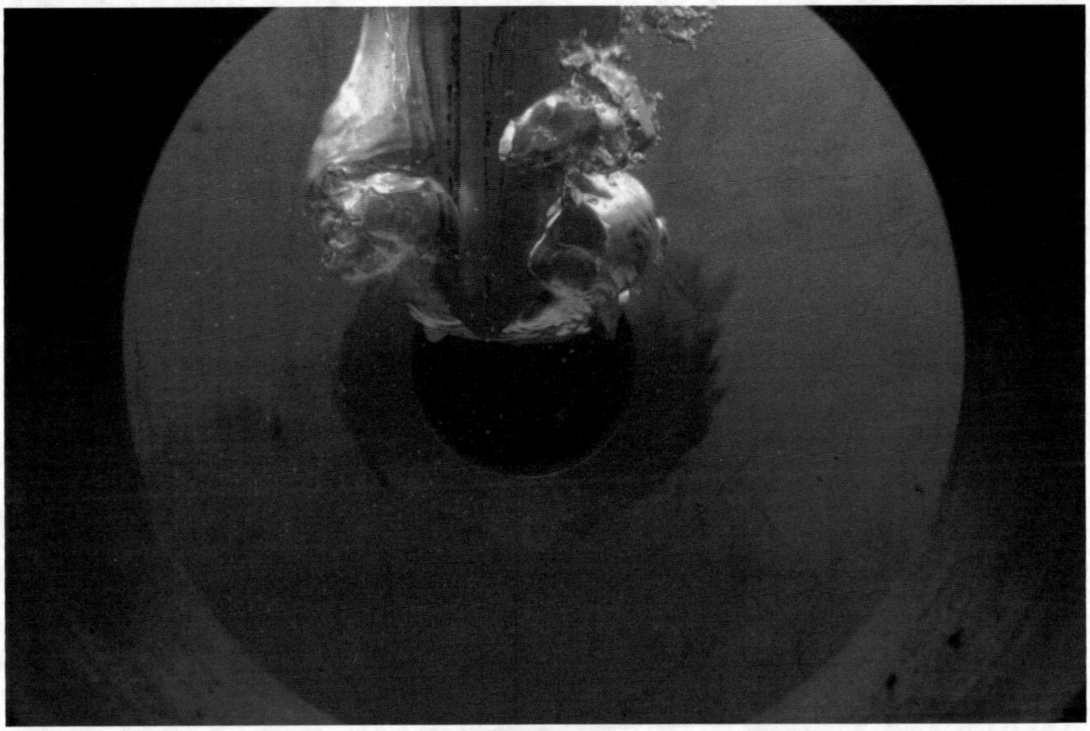

Figure 9: Bubbles Near the Column Bottom Viewed through a 4 Inch Nozzle

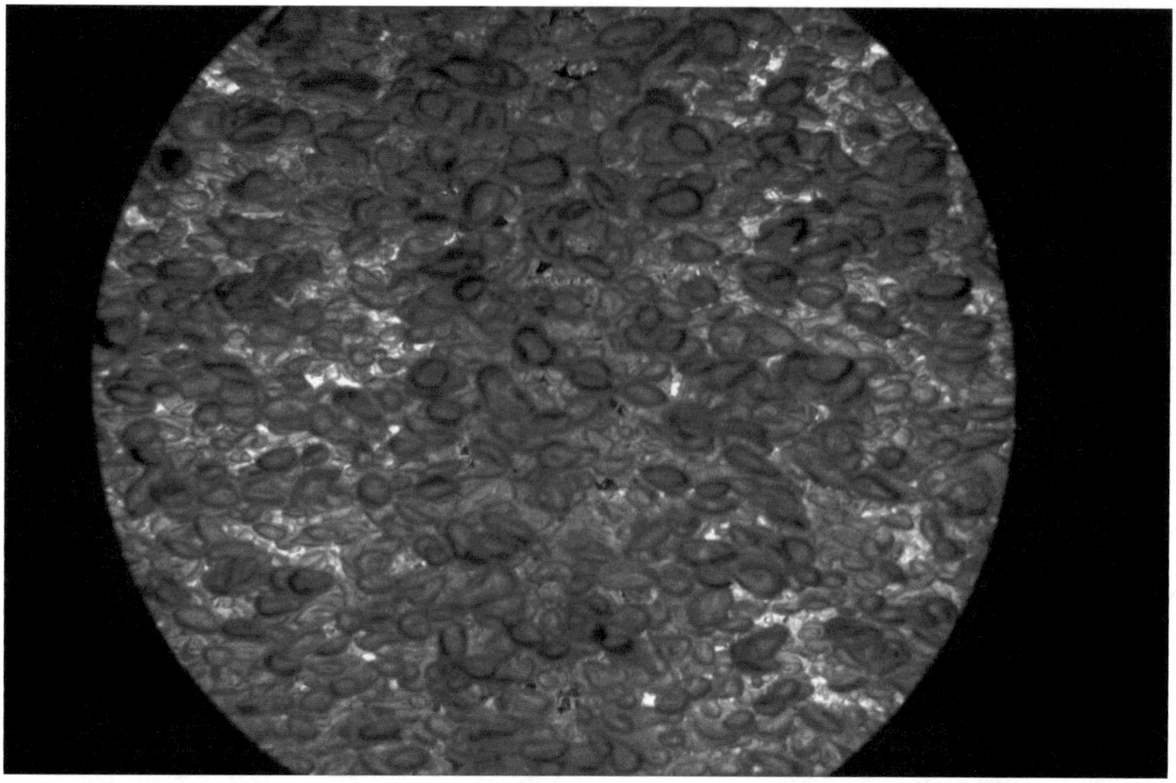

Figure 10: Bubbles Near the Column Top Viewed through a 4 Inch Nozzle

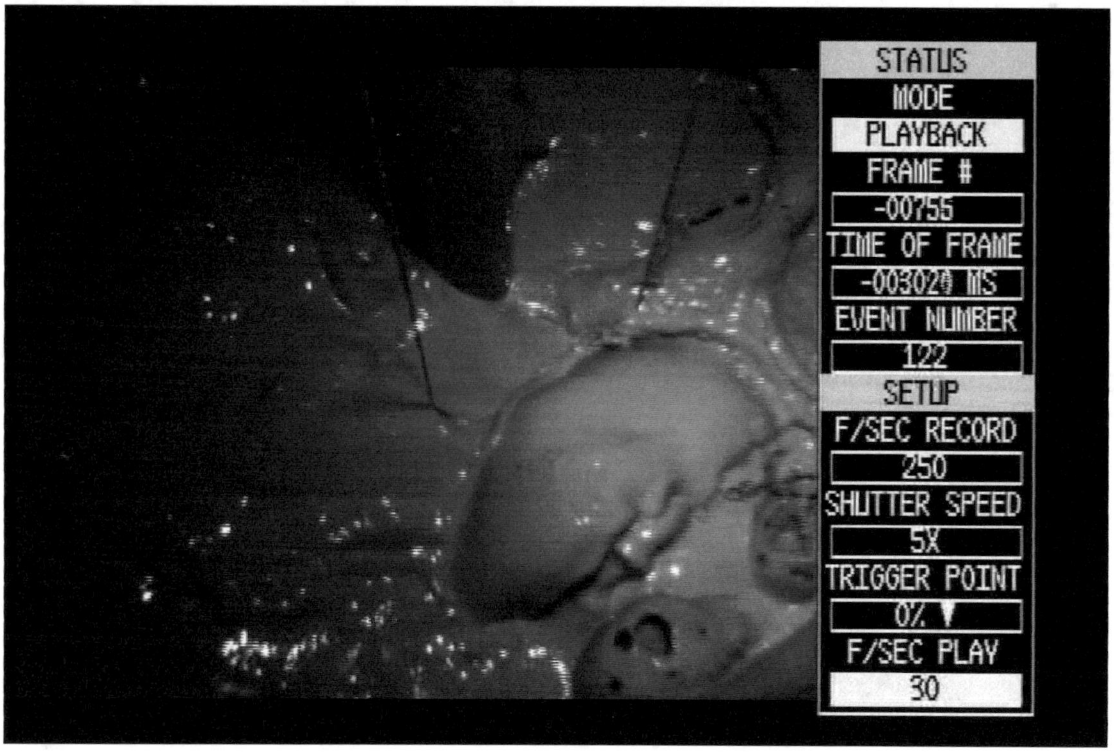

Figure 11: Bubbles on the Surface of the AZ101 Simulant

Void Fractions

Void fractions were measured with lasers (Figure 12) and pressure transducers as shown in figure 1. Typical laser test results are shown in Figure 13. Calculations were performed to determine the void fractions, φ, for each of the tests, using the gas volume, V_g, and the stimulant volume, V_s, such that

$$\phi = \frac{V_g}{V_g + V_s}$$

Figure 12: Laser Measurements for Water Tests

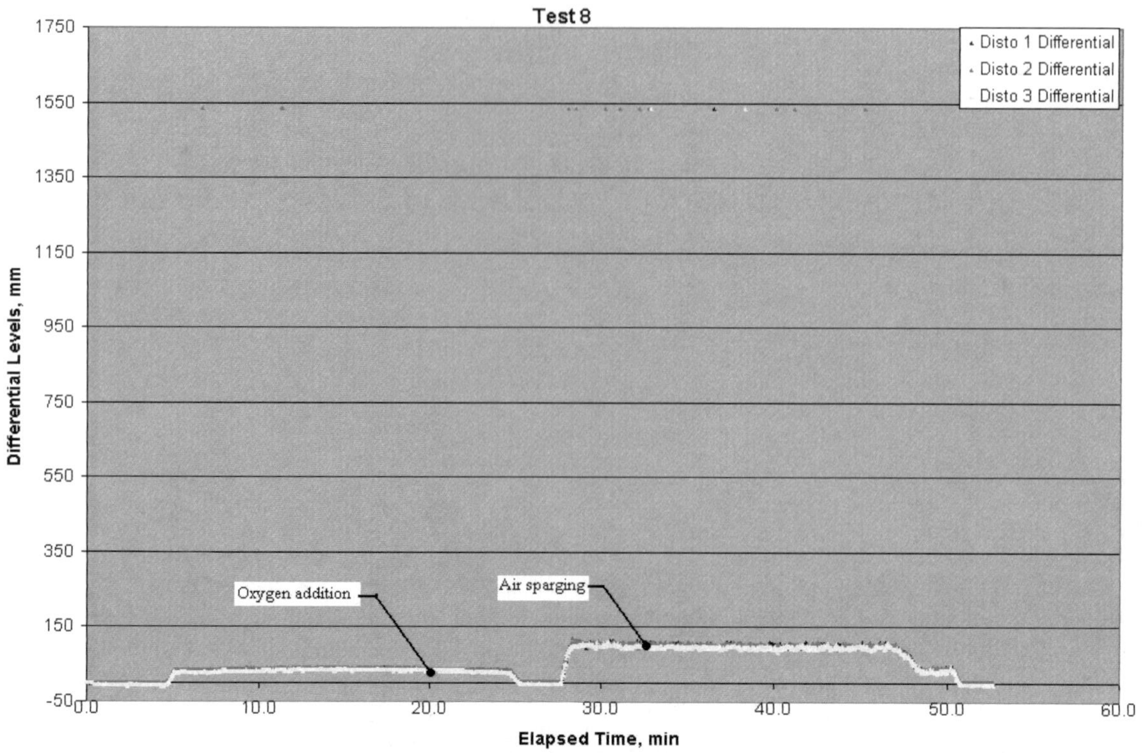

Figure 13: Laser Measurement Data

Test Results

All of the test results are displayed in Figures 14 – 21. Testing with water and water with AFA provided reasonable results, but the AZ101 testing raised concerns. In all testing, mixing at the lower levels of the column affected test results, due to bubble distribution and flow turbulence. For AZ101 tests, the void fraction uncertainty is affected by the way that the tests are performed. The calculated uncertainty showed that the maximum instrument uncertainty with respect to the void fraction is 0.5% for the water tests and 0.8% for the AZ101 tests. The values for water were reasonable, but the values for AZ101 were frequently outside of the measurement uncertainty. Following initial uncertainty calculations, a sample of the AZ101 was retested after mechanical agitation to determine if trapped air was present in the fluid. Testing showed that approximately 1.7 % of the volume was trapped air. The measured void fractions are actually changes in void fractions, rather than overall void fractions. This phenomenon affects any void fraction measurement in laboratory testing or elsewhere. Once air is sparged into solution, there will always be trapped air, unless mechanical agitation is introduced. The quantity of trapped air is variable. Bubbles are released from AZ101 solution after sparging has stopped, which further complicates the situation. In addition, as soon as sparging stops, the simulant begins to settle rapidly. For the tests performed here, the simulant was mixed to obtain homogeneity immediately prior to test, but DO concentrations near the tank bottom are inconsistent with typical results, when Figure 8 is compared Figure 22. Actual field results may vary for cases where settling has occurred prior to sparging. For example, in 60 days the yield stress of the material tripled during storage, which implies that additional gas holdup may initially exist in radioactive environments prior to sparging, since a higher yield stress material will hold up more gas. In short, trapped air affects void fraction results.

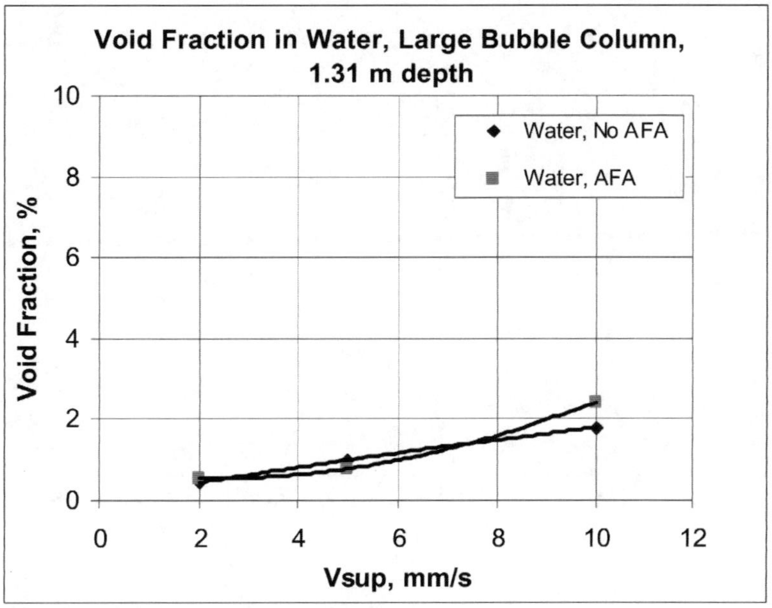

Figure 14: Average Void Fractions for Water without AFA Tests at 1.31 Meters

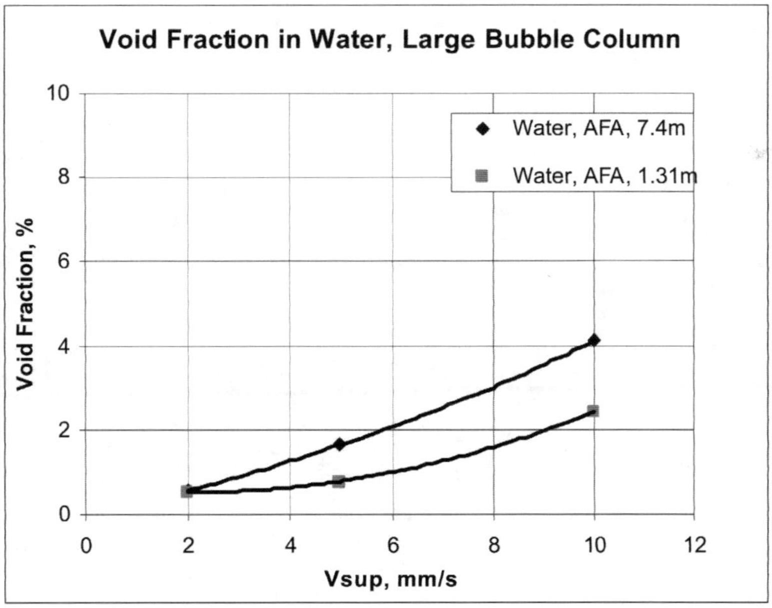

Figure 15: Comparison of Void Fractions for Water with AFA Tests

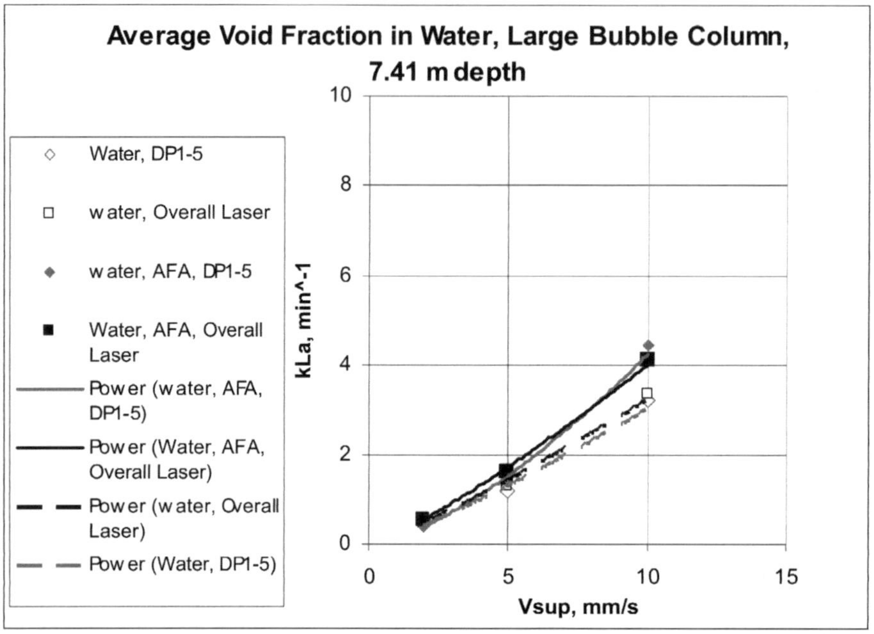

Figure 16: Average Void Fraction for Tests at 7.41 Meters

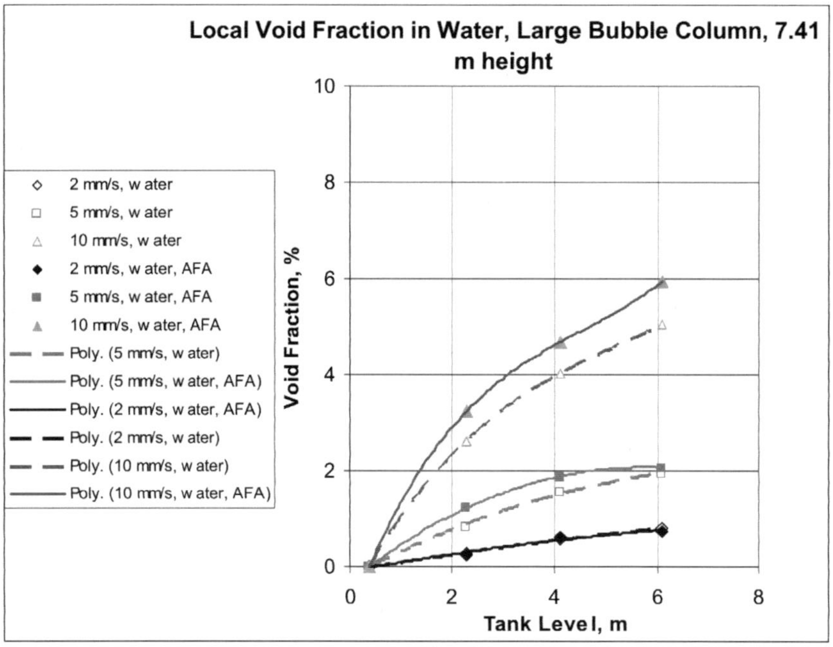

Figure 17: Local Void Fraction for Tests at 7.41 Meters

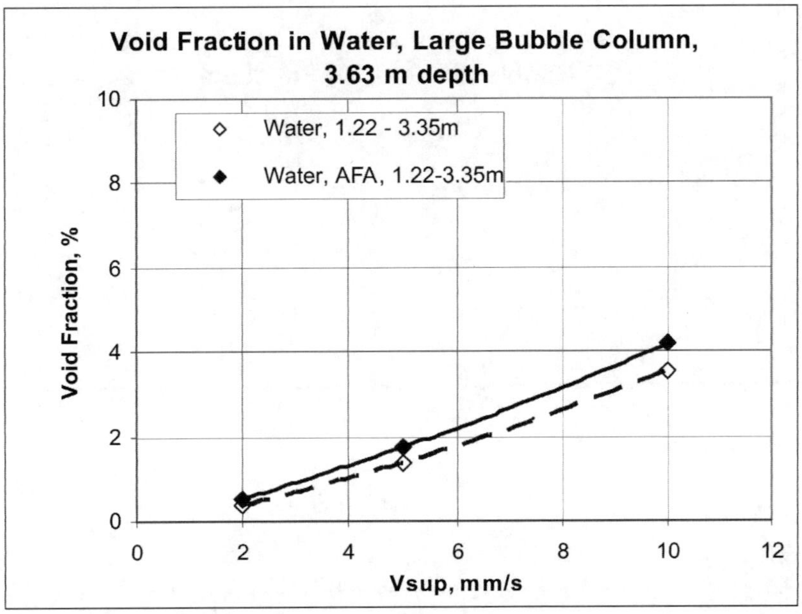

Figure 18: Local Void Fractions for Tests at 3.63 Meters

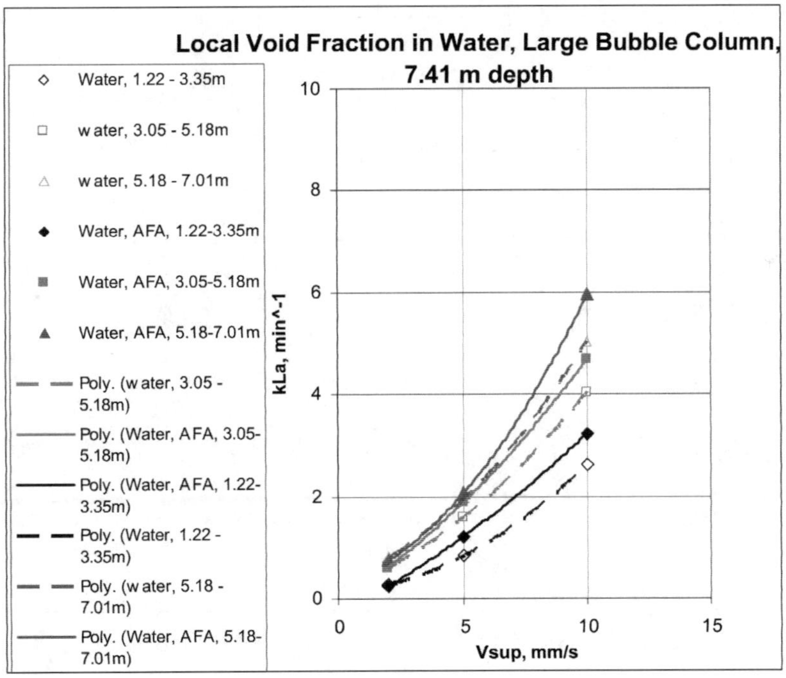

Figure 19: Local Void Fractions for Tests at 7.41 Meters

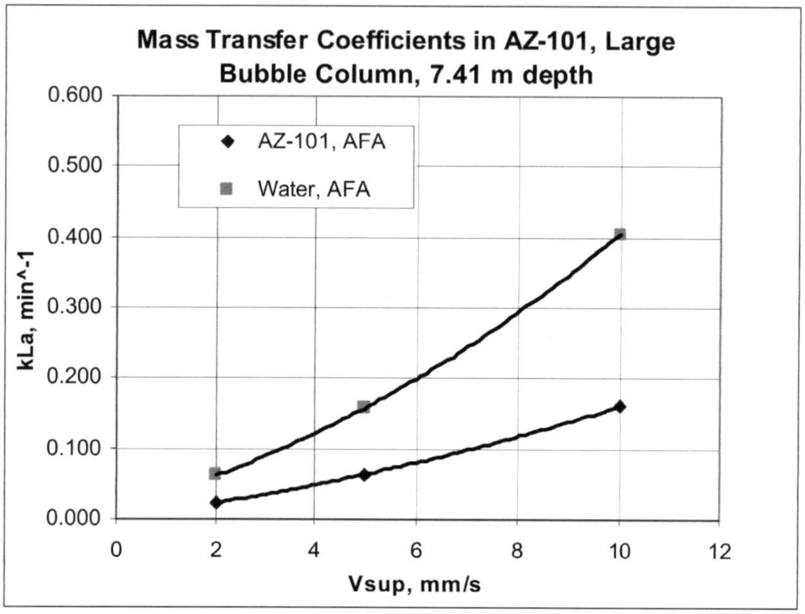

Figure 20: Local Void Fractions for AZ101 Tests at 7.41 Meters

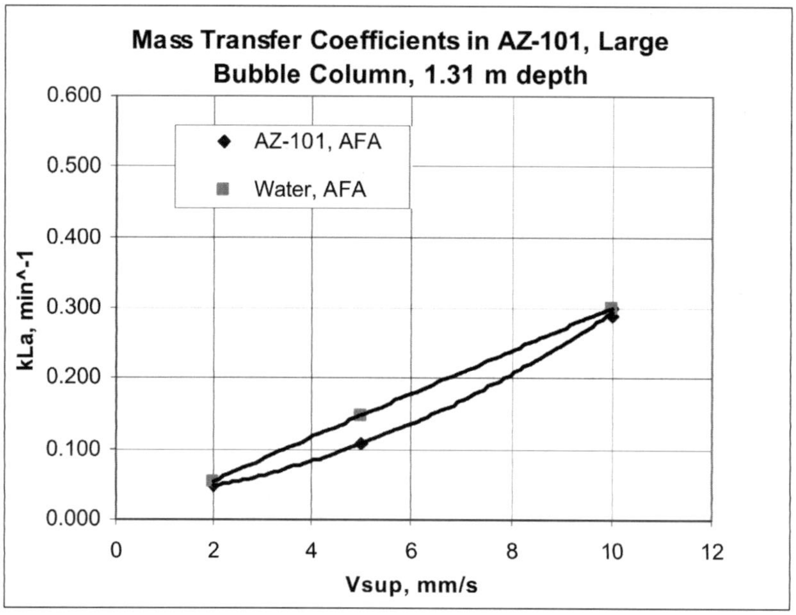

Figure 21: Local Void Fractions for AZ101 Tests at 1.31 Meters

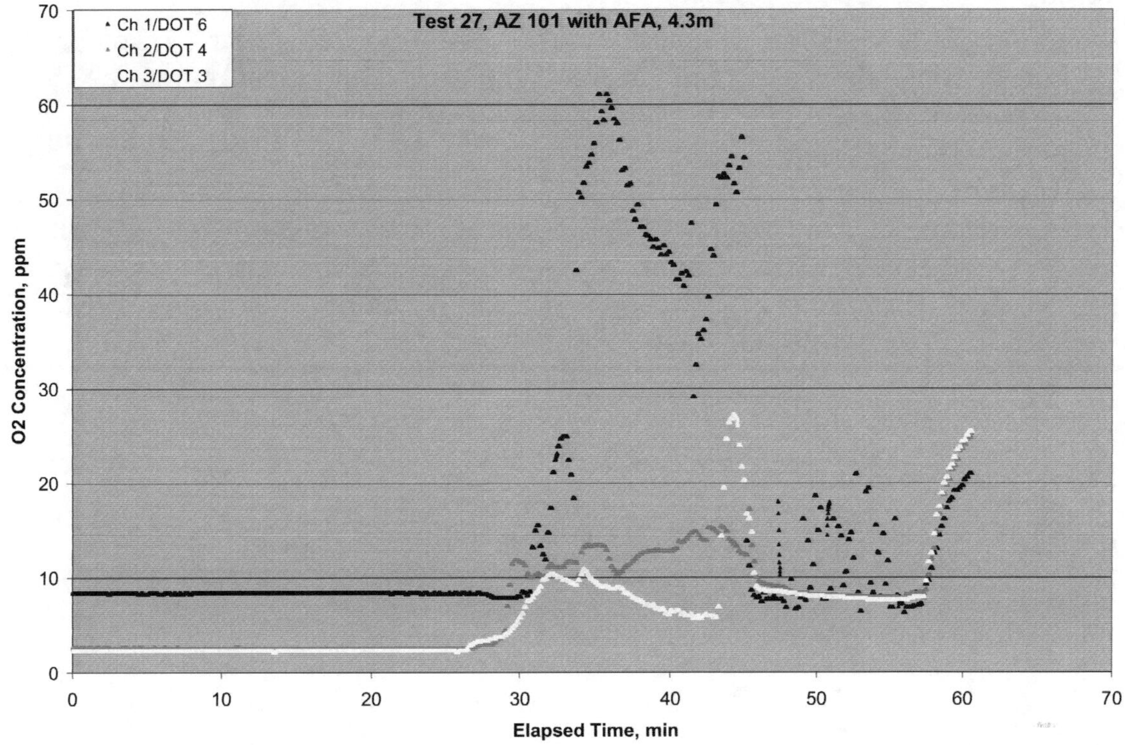

Figure 22: DO Concentrations Near the Column Bottom

Conclusions

This paper focused on mixing and void fraction results for large scale mass transfer tests, which were performed on several simulants including process water, process water with anti-foam agent (AFA), and AZ101 simulant with AFA. To do so, a 28 foot tall by 30 inch diameter, stainless steel column was constructed, along with supporting equipment to perform the required tests. The tests required simulant testing at various levels in the column, and consisted of saturating the simulants with oxygen. Oxygen was injected into the system, rather than generated in situ using chemical reactions to create distributed oxygen throughout the solution. Once saturation was obtained, a vertical, centrally located sparge tube injected air into solution at the bottom of the column to obtain specified superficial velocities, which are the average velocity across the simulant surface in the column. The injected air displaced the oxygen from solution, and the effects of this process were measured. In particular, the effects of mixing on the determination of void fractions were considered, and void fractions were shown to be markedly affected if air was used to mix a Bingham fluid prior to test, since trapped air remained in solution between tests. Also, poor mixing near the bottom of the column was evidenced by poor mixing from the air sparger, fast settling of solids in solution, and trapped air bubbles in solution. This paper concludes a series of three papers, which considered mixing, bubble formation, and mass transfer in a large scale system.

References

1. Leishear, R. A., Restivo, M., Guerrero, H., Sherwood, D., “Mass Transfer Coefficients for a Bingham Fluid and Water With and Without Anti-foam Agents”, 2008 Pressure Vessel and Piping Conference, ASME, N. Y.
2. Leishear, R. A., Restivo, M., Sherwood, D. “Bubble Formation in a Large Scale System”, FEDSM2008-55013, 2008 Fluids Conference, ASME, New York
3. H. N. Guerrero, C. L. Crawford, M. D. Fowley, R. A. Leishear, M. L. Restivo, “Effects of Alternate Antifoam Agents, Noble Metals, Mixing Systems, and Mass Transfer on Gas Holdup and Release from Non-Newtonian Slurries”, WSRC-TR-2007-00537, Savannah River National Laboratory, S. C.

301c Observations of Slurry Concentration Uniformity on Draining a Mechanically Agitated Vessel

Matthew Hunnemeder[1], **Robert P. Hesketh**[1], Bethany Schmid[1], YeJI Shin[1], and Arthur W. Etchells III[2]. (1) Chemical Engineering, Rowan University, 201 Mullica Hill Road, Glassboro, NJ 08028, (2) Industrial Mixing, Philadelphia, PA 19106

Draining slurries containing settling solids is common in industrial operations. In many situations the delivery of slurry with a uniform concentration is desired. This slurry delivery configuration is a complex unsteady state multi-phase problem and has not been the subject of intense research. A laboratory scale reaction vessel drained by siphon and mixed with a high-efficiency A310 impeller was used in this study. In order to determine the effect of agitation on the system, impeller speeds were investigated on a range of 0-600 RPM for a preliminary model using just water. For the range investigated, it was found that agitator speed has no effect on gravity flow out of the vessel as long as no air is entrained. Following this, tests were conducted in which the outlet concentration of sand-water slurry was measured as a function of draining time and slurry level in the stirred tank. Impeller speeds of 900-1500 RPM were used for the slurry delivery system charged with 10% solids by mass. A Mettler-Toledo Lasentec probe was used to determine the solids distribution in a stirred tank. From this, it was concluded that using a higher agitation speed increases the uniformity of the slurry efflux concentration. Impeller speeds of 1200-1500 RPM delivered more uniform slurry. With the Lasentec data it was determined that there was a maximum slurry concentration in a region above the impeller. This region was approximately 0.7 impeller diameters measured vertically from the bottom of the impeller. It was also observed that this high slurry concentration is independent of the liquid height in the tank.

301d Oil Viscosity Impact on the Production of Pickering Emulsions

Louis Fradette, Charles-Olivier Fournier, and Philippe Tanguy. URPEI, Chemical Engineering Department, Ecole Polytechnique de Montreal, P.O. Box 6079, Station Centre-Ville, Montreal, QC H3C3A7, Canada

Pipeline transportation technologies for viscous oils are becoming increasingly important because of the limited upgrading capacity in remote areas and the prohibitive cost related to building additional capacity. Heavy oil emulsification offers an attractive alternative to dilution for long-distance pipeline transportation. Pickering emulsions (solid-stabilized emulsions) could offer additional benefits when compared to standard emulsification practices with surfactants. Relying on the known and controlled wetting behavior of iron particles, the role of the dispersed phase viscosity in solid stabilized oil-in-water emulsions was studied in this work. Silicon oils of viscosities ranging from water-like to one Pa-s were used in a standardized emulsification process. Emulsification in controlled conditions showed that the viscosity of the dispersed phase dampened the particle anchoring at the o/w interface, slowing down the emulsion stabilization. The role of the dispersed phase viscosity is hence a key design factor for a continuous emulsification process. Based on the emulsification results, a combination of dispersed phase viscosity and agitation time must be respected in order to produce a stable emulsion. A relation between the emulsified oil volume, the o/w viscosity ratio, mixing power density, and agitation time is proposed for the studied systems.

Studies on Impeller Design on Power Consumption from Temperature Rise Data In a Stirred Tank

Damaraju P. Rao and Chander S. Mahey. Chemical Engineering, Indian Institute of Technology, Delhi, Hauz Khas, New Delhi, India

The power input to a mixer is dependent on the impeller type, its rotational speed, the physical characteristics of the processed materials, the shape of the tank and the relative locations of its components such as the baffles, impeller etc. and more importantly the amount of dispersed phase fraction such as solids or an immiscible liquid phase. Reduction in the power consumption is a key requirement of the industry to make it competitive and cost effective. In the present work, it was proposed to study power input measurement from temperature rise data and also study the mixing and power input performance of the modified hydrofoil impellers made by CME engineers to reduce power consumption by creating new impellers which are the modified versions of Hydrofoil impellers obtained by twisting 3 Bladed Hydrofoil impellers. For narrow blade hydrofoils, Power No is low for hydrofoils compared to other impellers. Typically, low power number (0.1–0.5) impellers generate mean flow whereas high power number impellers (>3) generate flow having more turbulent kinetic energy. Accurate estimation of the power input required to obtain the desired performance of mixing equipment is crucial for successful design of the mixing equipment., **Panja and Rao** (1) investigated a new technique to determine the power input from measured data on rate of temperature rise due to the stirring of a liquid or a gas-liquid system in a 16 cm diameter mechanically agitated tank having standard tank configuration.. The change in the temperature due to stirring was continuously measured by using a thermopile consisting of two hot and two cold junctions and the power input was calculated from known heat capacities of the liquid, the reactor vessel and the stirrer and the measured rate of temperature rise by using the following energy balance equation

$$(m_w C_{pw} + m_r C_{pr} + m_s C_{ps})\Delta T / t = P \tag{1}$$

where

m_w=Mass of water
C_{pw} =Specific heat of Water
m_r= Mass of wetted reactor
C_{pr} =Specific heat of reactor
m_s= Mass of stirrer+shaft+shaft support+ Baffle
C_{ps} = Specific heat of stirrer
DT = Temperature rise in ^{0}C in a time interval , t sec.

1. In the above equation, it is assumed that there are no heat losses from the reactor to the surroundings and also that reactor body, stirrer and the reactor fluid are at the same temperature as that of the reactor. The above assumptions are best realized if the initial temperature of the reactor is chosen same as the temperature of the

surroundings and small temperature rises can be accurately measured and if the rate of heat losses are small compared to rate of heat generated due to stirring and heat capacity of liquid is large compared to that of the walls. In addition to applying the temperature rise method, it was also proposed to measure electrical power input with and without water taken in the tank from the display screen of AC drive by pressing the up and down key and estimate mechanical power input to the tank..

DESCRIPTION OF THE EXPERIMENTAL SETUP AND PROCEDURE

.

A complete sketch of the experimental setup is given in Figure 1 and a brief discussion of the various components that are used in the setup are given in Table 1

- The experiments in the present work have been carried out with an agitated cylindrical vessel made of Perspex. The vessel is equipped with an opening in the top cover for charging the required quantity of liquid and/or solids. The height of the liquid level in the tank is kept at 49 cm. Soft water is used in the power consumption studies . There are four baffles inside the tank to avoid any vortex formation while operating at high velocities. A drain at the bottom corner is provided to facilitate easy draining of the products.
- The measurements were carried out within the turbulent regime of the fluid flow in the vessel (*Re 5**10^5 to 1*10^6).
- RO water at temperatures around 25°C was used. The temperature rise was measured with the help of 4/4junctions thermopile. The agitator speed was set by means of AC drive provided by Dan Foss.
- Runs were made at different stirring rates of a blade with and without solids (PVC granules) for measuring power consumption .

Experimental Procedure for the determination of Power Consumption

2. The tank is filled with water by using a water pipe through the opening at the top of the vessel. The water level in the tank is maintained equal to the diameter of the tank. To study the effect of solids, PVC granules of required weight are added in the water taken in the tank for making a predefined % solids loading.
3. After the tank has been filled up to the required level, the temperature of the water in the vessel is recorded by using a thermopile consisting of four hot junctions and four cold junctions and a digital micro-voltmeter connected to the thermopile wires,.
4. The reference junction of the thermocouple is inserted into the water bath which is maintained at constant temperature and the other is inserted into the tank.
5. The value of voltage developed by the thermopile is recorded at fixed interval of time
6. The stirring is continued at a fixed number of rotations per minute for a required length of time. The regulator knob in the VFD drive is rotated to get the required number of rotations per minute of the shaft. The number of rotations per minute is displayed on the display screen of drive and is noted down.
7. During the run it must be ensured that the variation in the number of rotations per minute should not deviate by more than 2%.
8. At the end of a run, the motor is switched off and either water is allowed to cool or is drained and replaced.
9. The electrical power inputs with and without water taken in the tank are also measured by noting the value directly from the display screen of AC drive by pressing the up and down key.

10. The above steps are repeated at different stirring speeds in steps of 50 rpm and for different % of solids loading.

Table. 1 Design details of the tank

<table>
<tr><th>Equipment</th><th colspan="2">Specification</th></tr>
<tr><td>Tank(Perspex)</td><td colspan="2">ID=490mm
OD=500mm
Height=910mm</td></tr>
<tr><td>Solid Shaft(SS316)</td><td colspan="2">OD=25mm
Length=800mm
Weight=4kg/m</td></tr>
<tr><td>Shaft support(SS316)</td><td colspan="2">OD=50mm
ID=44mm
Length=550mm
Weight=3.11kg/m</td></tr>
<tr><td>Impeller(SS316)</td><td colspan="2">Hydrofoil impeller(Blade1,2,3,4)</td></tr>
<tr><td>4 Baffles(SS316)</td><td colspan="2">Thickness =3mm
Length=725mm
Width 450mm</td></tr>
<tr><td>Stainless steel lid</td><td colspan="2">Provided with water inlet, thermocouple inlet and conductivity probe inlet</td></tr>
<tr><td>Tank Stand</td><td colspan="2">Circular stand of height =10.5 cm</td></tr>
<tr><td>Electric Motor</td><td colspan="2">Hindustan electric motors,1 HP,3P, 415 V,2A,75% efficiency</td></tr>
<tr><td>Motor controller drive</td><td colspan="2">Danfoss, VLT 2800 series</td></tr>
<tr><td>Conductivity Meter</td><td colspan="2">Global DCM900</td></tr>
<tr><td>Conductivity cell</td><td colspan="2">Cell cons.=1.0</td></tr>
<tr><td>Digital multimeter</td><td colspan="2">Sanwa,PC5000</td></tr>
<tr><td rowspan="2">Thermocouple</td><td rowspan="2">Constantan
Chromel (E type)</td><td>2 junction thermocouple supplied by K.M.Instruments,sensitivity=0.05263mV/C</td></tr>
<tr><td>4 junction thermopile made in our lab, sensitivity=0.1265mV/C</td></tr>
<tr><td>Water bath</td><td colspan="2">Julabo,Model ME</td></tr>
<tr><td>Temperature controller</td><td colspan="2">Audiotronics (with RTD Probe)</td></tr>
<tr><td>Humidity meter</td><td colspan="2">Instron 1003</td></tr>
<tr><td>Personal Computer and UPS</td><td colspan="2"></td></tr>
</table>

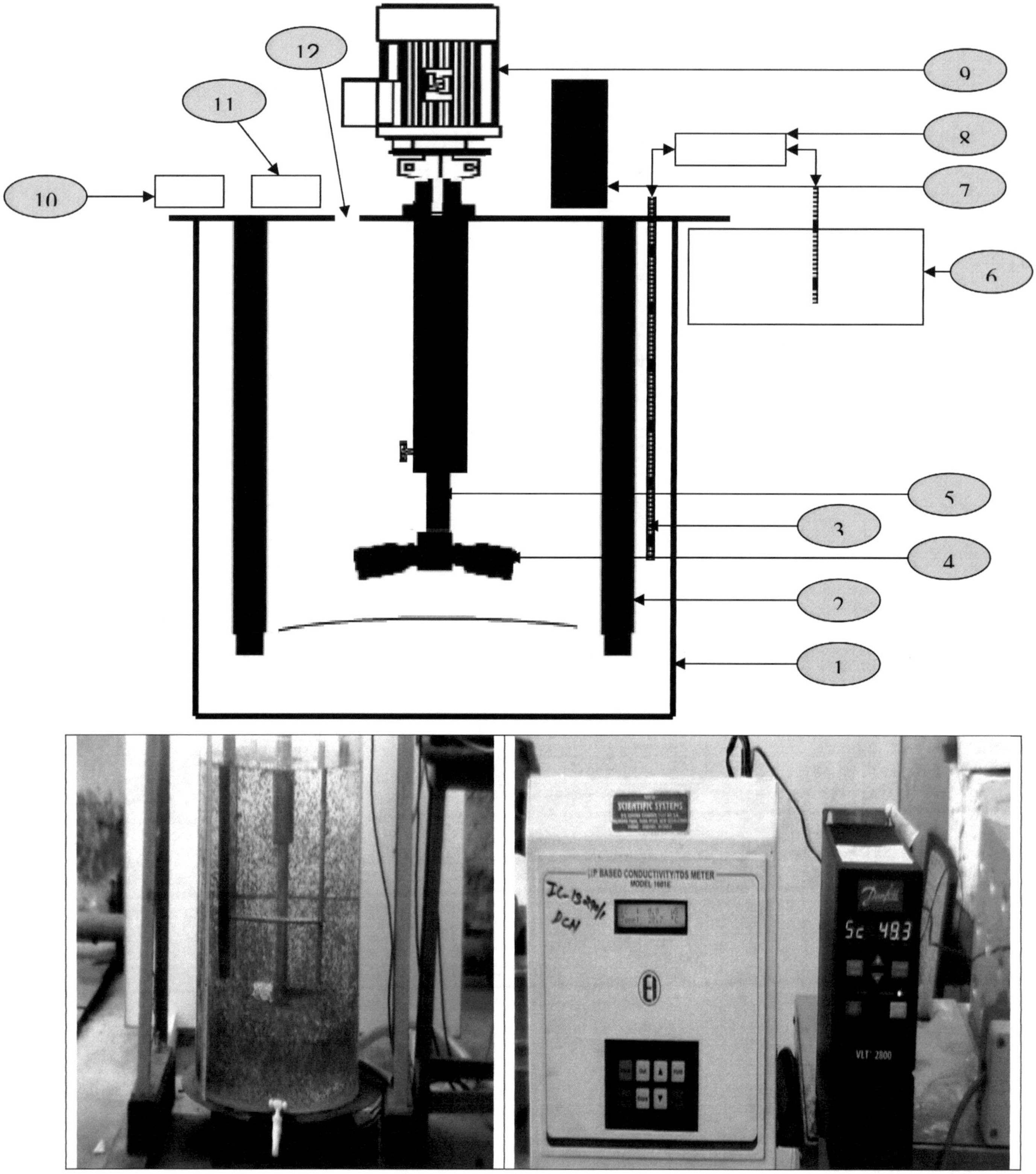

1.Flat Bottom Tank, 2.Baffles, 3. A 4 junction Thermopile, 4.Hydrofoil Agitator, 5. Motor Shaft, 6. Water Bath, 7. Motor Controller drive, 8. Digital Multimeter, 9. 3P Electric Motor, 10. Temp. Controller, 11. Humidity Meter, 12. Inlet port

Figure .1. Schematic diagram of Experimental Setup

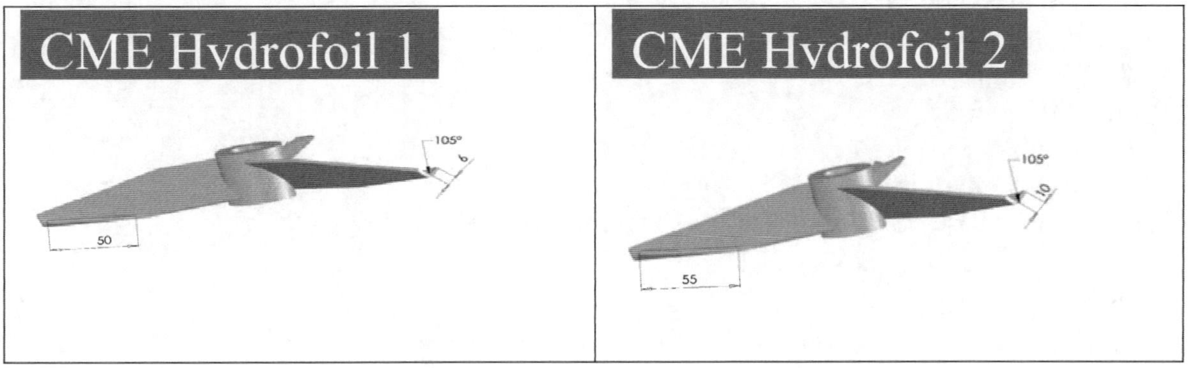

Figure 2. Impellers used for study

.Calculation of power input from rate of temperature rise of tank fluid was done by using the following relation:

$$P=(m_w C_{pw} + m_r C_{pr} + m_s C_{ps})\Delta T/t$$

where
Mass of water=92.128kg
Cp =4.1855 KJ/(KgC)
Mass of wetted reactor=7.22 kg
Cpr =1.27 KJ/(KgC)
Mass of stirrer+shaft+shaft support+ Baffle=0.4+3.2+1.71+3.5=8.81kg
Cps =0.494 KJ/(KgC)
P=399.123*ΔT/time, Watts
Figure 3 shows a typical run showing thermocouple emf recorded as a function of time which shows that raise in tank temperature is linear with respect to time.

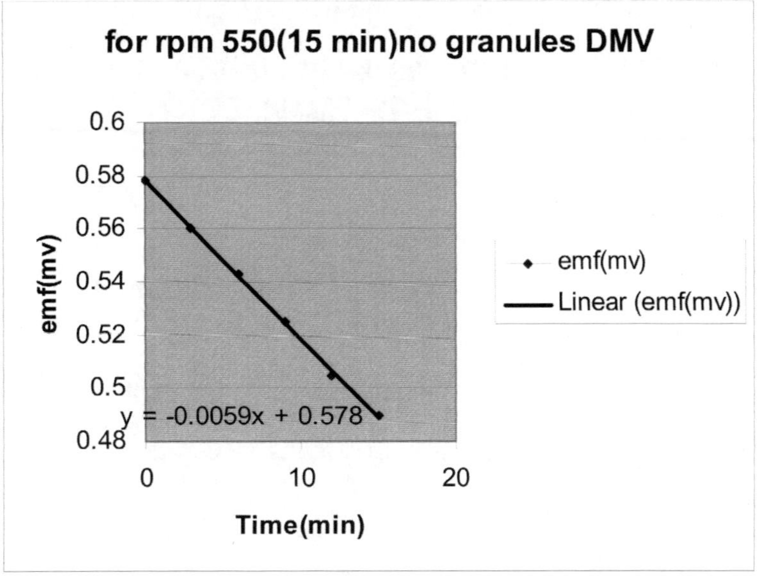

Figure 3. Thermopile voltage as a function of time for blade No. 1

The sample calculations for power input for the data shown in Figure 3 are given below:

Sensitivity of thermopile = 0.268 mV/C
Total run time =15 min
Reference temp.=32.5°C
Rpm=550
Calculation:
Slope=0.0059mv/min
Rate of temp. Rise= Slope (mV/min)/ sensitivity of thermopile (mV/^{O}C)
=.3.669×10^{-4}°C/sec
P=$399.123\times10^{3}\times 3.669\times10^{-4}$
=146.4W
Power consumed by motor with water inside the tank (AC drive) =230W
Power consumed by motor without water inside the tank =70W
Net electrical power input corrected for electrical and mechanical energy losses in the motor=160W

Accuracy analysis

Error in measurement in temp rise method

e1 = error in measuring initial temp T1 in terms of thermoemf in μV

e2 = error in measuring final temp T2 in terms of thermoemf in μV

Max error = [(|e1|+|e2|)/(/Δe)]X 100%

Least count of Microvolt meter = 1μV

Δe = Measured change in thermopile voltage over a time interval t sec..

The relative error in power input voltage change over an interval of time has been found to be between 2% to 10% .

Heat Capacity from Perspex tank & undipped part of stirrer in air.

= ((WCp)perspex tank +(WCp)hanging stirrer))/WCp (total)

=((7.22X1.27)+(4 X 0.494)/399.123 X 100%

=2.79%

Error of measurement in Wattmeter

LC = 10 w , hence error could be [10/P] x 100%. Hence, the method of power input measurement by using wattmeter is reliable for P>100 W.

Tables 2-4 show the power input measured by using the method of temperature rise and net electrical power drawn as a function of stirring rate obtained by using the three impellers.

Table 2. Power Measurement for CME hydrofoil impeller 1

RPM	Time (Min.)	Method 1 (Net electrical power consumed in mixing) in Watts	Method 2 (Power from temperature rise data) in watts
350	15	40	25.1
400	15	60	49.89
450	15	90	99.7
500	15	130	119.73
550	15	160	146.4
600	15	190	182.43

Table 3.Power Measurement for CME hydrofoil impeller 2

RPM	Time (Min.)	Method 1 (Net electrical power consumed in mixing) in Watts	Method 2 (Power from temperature rise data) in watts
350	15	30	27.30
400	15	60	54.60
450	15	90	84.39
500	15	100	89.35
550	15	150	136.51
600	15	180	166.33

Table 4. Power Measurement for PTD impeller

RPM	Time (Min.)	Method 1 (Net electrical power consumed in mixing) in Watts	Method 2 (Power from temperature rise data) in watts
350	15	100	91.8
400	15	140	124.11
450	15	200	178.71
500	15	290	275.39
550	15	380	374.79
600	15	490	474.03

Study on Effect of twist in hydrofoil impeller

A Comparison of power drawn found by using the temperature rise method for the three impellers corresponding to three agitation rates for water as the tank fluid is shown in Table 5

Table 5

RPM	500	550	600
IMP 1	120.0	146.4	182.4
IMP 2	89.4	136.5	166.3
PTD	275.4	374.8	474.0

Results were obtained by varying solid loading of PVC particles from 0.1% to 0.4% but as the solid loading was very small, no significant change in power consumption was found.

CONCLUSIONS:

- Increase in twist of Hydrofoil impeller reduces Power consumption
- Power Consumption data measured by temp rise method and estimated from net electrical power input consumption were in reasonable agreement
- Power consumed was found to be increasing with increase in rpm.
- The Difference in the power input data obtained by using the two methods was of the order of 3% for higher rpm to 18% for lower rpm i.e the difference was found to be decreasing with increased power consumption
- Power Consumption for Hydrofoils having greater twist was lesser than Hydrofoils having lesser twist.
- Power consumption for CME hydrofoil blades was lesser than Standard Pitched blade turbine impellers
- The solid loading being very small as such , the change in power consumption due to solid loading could not be detected by the power meter

In the case of the temperature rise method, the power consumption data as a function of sold loading did not show any expected trend due to the following reasons:

:

1. The heat capacity of solids is not added in the power consumption calculation. However, for low loadings , contribution by heat capacity of solids will be small.
2. But as the solids are thermally poor conducting type, there may be temperature gradients in the solids, and heat transfer from liquid to solid phase under transient conditions may be significant and hence the temperature probe may be registering a lower rate of rise with increase in solids. Hence further work is needed to modify the method of power input by temperature. rise method for solid -liquid systems.

REFERENCE

[1] N.C. Panja and D.P. Rao, Power input measurement from temperature rise data in mechanically agitated contractors, *Institution of Chemical Engineers*, (1992) 24-27.

434b Improving on a Good Thing: The Chemineer Xe-3 High-Efficiency Impeller

Eric E. Janz, Chemineer, Inc., P.O. Box 1123, Dayton, OH 45401, Julian B. Fasano, Mixer Engineering Company, 2673 Stonebridge Drive, Troy, OH 45373, and Kevin J. Myers, University of Dayton, 300 College Park, Dayton, OH 45469-3433.

The introduction of low-power number axial-flow impellers revolutionized the design of agitators for flow-controlled processes such as blending, heat transfer, and solids suspension. These impellers were capable of meeting process objectives at lower torque and power levels than traditional pitched-blade turbines. The Chemineer HE-3 is an example of a first generation high-efficiency impeller that has found numerous applications in the chemical processing and allied industries. Building upon the success of the HE-3, a new high-efficiency impeller, the XE-3, has been designed to further improve impeller performance, both in terms of mechanical characteristics and process performance. By virtue of their shape, the blades of the XE-3 are stronger than those of the HE-3, thus reducing the weight of the impeller. This provides a number of benefits, including lower material costs and the possibility of using higher speeds, and thus lower torques, and/or longer shafts. Although these mechanical improvements are quite positive, obtaining them at the cost of reduced process performance would not be acceptable. Thus, the XE-3 is designed to provide higher pumping efficiencies and improved solids suspension performance than other high-efficiency impellers. This presentation will describe how the XE-3 simultaneously improves mechanical and process performance, including comparisons of the pumping efficiency and solids suspension performance of the XE-3, HE-3, and pitched-blade impellers.

2008 AIChE Annual Meeting
Philadelphia, PA
November 16 – 21, 2008

New Impeller for Gas Dispersion

Bob Higbee[1], Binxin Wu[1] and Wojtek Wyczalkowski[1]

[1]Philadelphia Mixing Solutions, Ltd., 1221 E. Main Street, Palmyra, PA 17078, USA

ABSTRACT

This project was initialized as a response to industry demand for an improved impeller for gas dispersion. Several impeller prototypes were built and tested in 0.61 m diameter test rig, CFD and PIV. Analysis was focused on different combinations of blade shape, blade size, and blade location in relation to axis of rotation. All impeller prototypes were made by Laser Stereo Lithography method to ensure accuracy of shape. The project resulted in an optimized design that provides excellent process results, can be built for high power applications, and can be assembled on site.

INTRODUCTION

Production of PTA, hydrogenation and fermentation processes require efficient gas dispersing impellers that can produce high mass transfer and have high gas handling capability therefore improving process economy. The scope of this work was to improve and optimize existing impeller design using a combination of experiments, CFD and PIV.

For decades the impeller of choice was Rushton turbine. The first break-through was by Prof. Smith who introduced a concave blade. Later, SCABA and John Middleton pioneered the sharp edge design on the trailing side of the concave blade. Both innovations helped to reduce drag coefficient of a standard Rushton blade.

When blades are rotating through a liquid the pressure in front of blade is lower than behind the blade. Gas phase introduced into the mix goes into large cavities in a streamline formed from the blades. The result is a drastic reduction of resistance and power required to turn the turbine. Power reduction deficit must then be adjusted by RPM increase, which is inconvenient for operation and requires specialized hardware. When concave blades are used, the effect of gas cavities is diminished and power decrease in gassed conditions is much lesser. Also, with smaller gas cavities the impeller is less prone to flooding.

It is generally desirable that a gas dispersing impeller has low Power Number that is constant or close to constant in gassed condition and can handle large amounts of gas without flooding. The point of flooding and gas handling characteristics strongly depend upon the blade shape, length and location in relation to axis of rotation. There is no literature data on systematic approach to gas dispersing impeller design.

Several prototypes of impellers were built and tested in order to determine how geometry influences mixing characteristics.

MODEL/EXPERIMENTAL

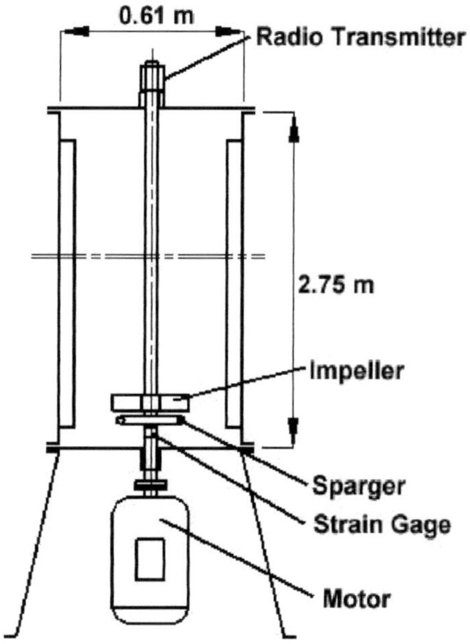

Fig 1. Experimental setup, 0.61 m diameter vessel x 1.83 m tall, Electric motor with Variable Speed Drive, Torque measurement with Strain Gages and Telemetry Transmitter. Air delivered through Toroidal Sparger

Tested Impellers:

Fig 2. Tested impellers (1)

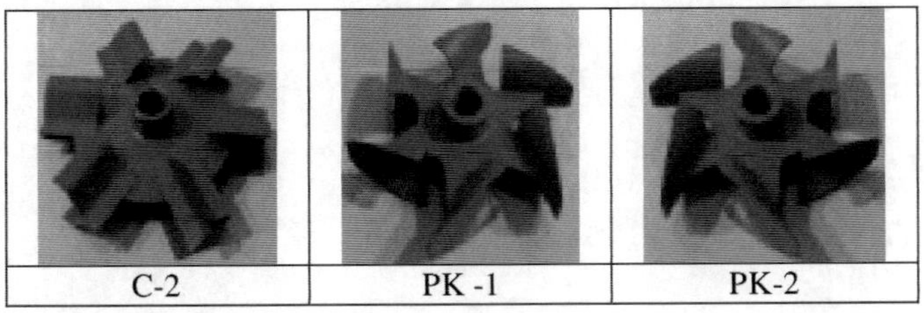

Fig 3. Tested Impellers (2)

Fig 4. Tested Impellers (3)

Impellers were tested for gas dispersion characteristics, solids suspension capability in gassed and un-gassed conditions. Impeller flow and dissipation were measured in 0.61 m diameter test vessel using PIV technique.

RESULTS

Selected test results are presented below

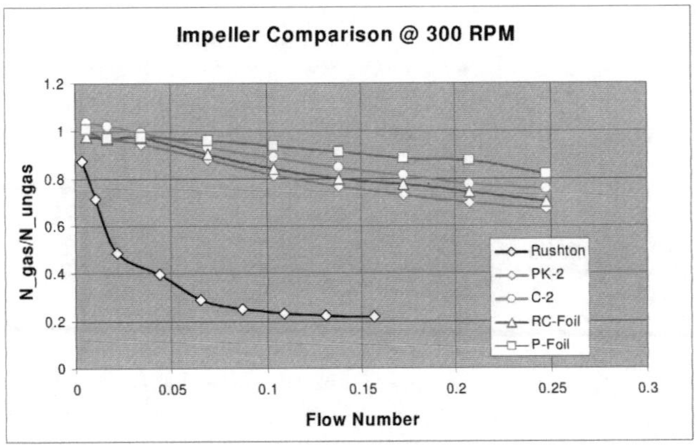

Fig 5. Comparison of selected impeller performance in gassed condition @ 300 RPM in 0.61 m diameter vessel.

Solids suspension capability was tested in 0.62 m diameter vessel with ASME dish bottom. All comparative tests were performed on sieved alumina particles with fraction size 125-149 microns. For each impeller Just Suspended speed was recorded. Experiments were done for gassed and un-gassed condition. Toroidal sparger was mounted under impellers.

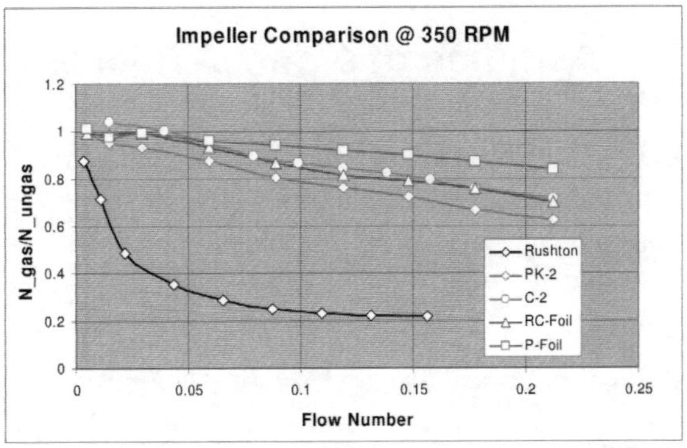

Fig 6. Comparison of selected impeller performance in gassed condition @ 350 RPM in 0.61 m diameter vessel

CONCLUSION

The final impeller design achieved desirable performance for gas dispersion as it has been presented in Fig 5 and 6. The gased power characteristics is flat. This design also performs very well in solids suspension applications.

Impeller blades are relatively easy to manufacture with a series of single bends.

Blade length significantly influences impeller performance. Blade length increase can be achieved by reduction of disk diameter. Elimination of disk does not have negative effect on impeller gassed characteristics.

REFERENCES

Cooke, M. and Heggs, P. (2005) “Advantages of the hollow (concave) turbine for multi-phase agitation under intense operating condition”, Chemical Engineering Science 60, (5529-5543)

Nienow, A. and Warmoeskerken, M,, Smith, J., Konno, M., “Flooding/Loading transition and the complete dispersal condition in aerated vessels agitated by a Rushton Turbine”, Proceedings of the Fifth European Mixing Conference, Germany, BHRA Fluid Engineering, Cranfield, pp. 153-154

Smith, J., van't Riet and Middleton, J (1977), “ Scale up of agitated gas-liquid reactors for mass transfer”, Second European Conference on Mixing, Cambridge UK, paper F4-51 t0 66

US Patent 4,779,990 (SCABA)

434d Agitation of a Shear-Thickening Mixture of Cornstarch and Water

Chad Michael Huelsman, Bradley J. Fischer, Joseph E. Paulsen, Michael J. Elsass, and Kevin J. Myers. Department of Chemical Engineering, University of Dayton, Kettering Labs Room 445, 300 College Park, Dayton, OH 45469-0246

The agitation of a shear-thickening cornstarch-water slurry was studied. The objective was to predict the effect of dilatancy on agitation and the desired operational settings for agitating dilatant materials of known composition. Mixtures of cornstarch in water from 45–52 wt. % were agitated using different impeller types, impeller sizes, and vessel sizes (8–12 in. diameter). Torque and rotational velocity data were recorded for each configuration at various operational settings. The rheology of each mixture was characterized by a Brookfield DV-II+ viscometer, which confirmed shear-thickening behavior. The shear-thickening behavior of each agitated mixture culminated in a point of system instability, a certain impeller speed under a given configuration at which the shear-thickening mixture would start to seize up and the shaft and impeller would deflect substantially. It was under this condition that plots of rotational speed and shear rate versus impeller-to-tank diameter ratio were constructed from the recorded data. Agitation performance at unstable shearing conditions was found to be a function of the ratio between the impeller diameter and the tank diameter and was relatively independent of the tank diameter and impeller type. This suggests that operational limits to avoid unstable shear-thickening behavior upon scale-up are predictable. Furthermore, a relatively linear relationship between the impeller rotational speed and the diameter ratio was observed at the point of instability. This linearity may also have implications for the scaling up of shear-thickening agitation. Further testing should be carried out to confirm these results and to develop from them a size-independent empirical correlation that can assist in the scale-up of dilatant agitation systems to achieve optimum performance without instability.

434e Mixer Performance and Scale-up Calculations Using a Microsoft Excel Based Tool

David A. R. Brown, BHR Group, The Fluid Engineering Centre, Cranfield, MK43 0AJ, United Kingdom

Through BHR Group's Fluid Mixing Processes (FMP) consortium, BHR have developed a novel Microsoft Excel based mixer performance calculation tool. This is essentially a correlation based tool, with an easy to use and familiar Excel based interface. A large number of proprietary and non-propriety mixer performance prediction functions and correlations have been implemented, covering the performance of single and muti-phase stirred tanks, in-line static mixer and jet mixed systems. The tool can be used to easily and quickly perform "what-if" and scale-up calculations. Other facilities included to aid use include an integrated unit conversion tool and a simple mixer diagram, as well as facilities to save and restore previous calculations.

The presentation will discuss the potential advantages and the sometimes significant risks associated with using this and other mixer performance calculation tools.

434f Using Positron Emission Particle Tracking (PEPT) to Study Mixing in Stirred Vessels

Mostafa Barigou, Paulina Pianko-Oprych, Luke Adams, Antonio Guida, Fabio Chiti, and **Alvin W. Nienow**. Department of Chemical Engineering, University of Birmingham, Edgbaston, Birmingham, B15 2TT, United Kingdom

Positron Emission Particle Tracking (PEPT) is unique in flow visualisation terms, being able to examine flow phenomena quantitatively in three dimensions in systems that cannot be studied by techniques such as PIV and LDA because they do not allow the passage of laser light. Slurries are a typical case. They are found in many industries including minerals, food processing, household and personal care products. Indeed, some 80% of products in the chemical industry, for example, are of this type. The suspension of solids in low viscosity fluids is one common example and is critical for promoting chemical reaction between the two phases, or for enhancing crystal growth. In the latter, a wide size range is found and establishing their spatial distribution as function of size has proved difficult though it is important. PEPT is able to do this and also determine the velocity of each of the two phases. In many other cases, especially in food processing, household and personal care products, slurries exhibit non-Newtonian, paste-like behaviour. Little work has been done to understand how to mix them effectively. The reason for this is their opaque nature. Therefore, it has been assumed that knowing the rheology of the slurry enables the prediction of its behaviour, based on studies on transparent polymer solutions with similar rheological properties. Again, PEPT can tackle such issues, giving a full flow field, viz., the three velocity vectors throughout the vessel. Such work has been carried out on paper pulp (a yield stress fluid), a personal care product (yield stress and viscoelastic) and on a china clay slurry (shear thickening}. All this work will be summarized, along with a critique of the technique and a study to establish its accuracy, based on a comparison with the extensive literature on Rushton turbines in water.

Novel Objective Functions for Static Mixers

E. Bruce Nauman
The Isermann Department of Chemical and Biological Engineering
Rensselaer Polytechnic Institute
Troy, NY 12180
Tel: 518 276-6726
Email: nauman@rpi.edu

Most static mixers are designed to maximize radial mixing to achieve compositional and thermal homogeneity. Flow inverters maximize heat transfer. Axial mixers that closely approximate an exponential distribution of residence times have been devised as well. More recent papers describe static mixers for surge dampening in industrial swing-cycle processes (Smith, Graham, and Palamara, 2006). An extension of this result allows complete attenuation of surges even when the component with periodic concentration variations is reacting by first order kinetics (Nauman and Smith, 2008) . Another class of motionless mixer is intended to maximize first appearance times in laminar flow, tubular reactors. Coiled tubes, especially with periodic changes in the direction of the principle axis have shown excellent results but are somewhat cumbersome in terms of spatial layout (Mridha, and Nigam, 2008). Baffles in a short, fat reactor used for sterilization have been shown effective. The general question is: what objectives functions are reasonable for static mixers and how can near optimal performance be achieved in practical hardware? The first question is addressed using the tools of residence time theory. The second question is usually addressed using commercial CFD packages, but these can be problematic. An alternative considered here is to simple, transparent code to test design concepts and to provide limiting cases for the verification of CFD code.

Laminar Flow Approximation to a Flat RTD

Following Smith, Graham, and Palamara (2006), a flat residence time distribution is

$$\begin{aligned} f(t) &= 0.5/t_p \text{ for } 0<t<t_p \\ &= 0 \text{ for } t>t_p \end{aligned} \qquad (1)$$

This distribution will completely dampen a concentration having period t_p. This statement is true for any waveform. The flat RTD also does an does an excellent of attenuating sinusoidal inputs of other periods and typically outperforms a CSTR of equal volume. The washout function corresponding to Equation 1 is

$$\begin{aligned} W(t) &= 1-t/t_p,\ t<t_p \\ &= 0,\ t>t_p \end{aligned} \qquad (2)$$

Note that $t_p = 2\bar{t}$ where $\bar{t}$ is the mean residence time. A previous paper (Nauman and Smith, 2008), showed how a flat RTD can be approximated by a shell-and-tube design when the tube flow is turbulent. The laminar flow case is actually a bit harder to do well. The straight line in Figure 1 is the desired $W(t)$ as given by Equation 2. The curved line in Figure 1 corresponds to five laminar tubes in parallel with equal flows per tube (20% of total). A numerical optimization, constrained by a constant $\bar{t}$, of tube volumes was based on sum-of-squares differences between actual and flat RTDs. A random optimizer was used, e.g. see Nauman 2008. Figure 2 shows the corresponding results for $f(t)$, the ideal curve being a flat line corresponding to Equation 1. Figure 3 shows the response of the five tube systems to sinusoidal disturbances. It outperforms a CSTR over a broad range of frequencies. Although not tested here, it is expected to perform better that a CSTR for disturbance of arbitrary waveform within this range of frequencies.

Maximization of First Appearance Times in Laminar Slit Flows

We consider the familiar parabolic velocity profiles for laminar flow in tubes and between flat plates. For a circular tube, the first appearance time is $t_{first} = \bar{t}/2$ and for flow between flat plates it is $t_{first} = 2\bar{t}/3$. For processes such a sterilizations, this makes a substantial difference in reactor volume and in the thermal time distribution (Nauman, 1977) and accounts for the common use of spiral and plate-type heat exchangers for processing heat sensitive materials. Figure 4 shows two s of a two-channel, slit flow geometry in which the relative channel widths can be varied. The velocity profile is parabolic in each , the pressure gradient down each channel is identical, and thus the mean velocities in the channels will be different except when the central baffle is located exactly at the center line, here denoted as $y = 0.5$. When there is only one , the optimal location for the baffle is at $y = 0.5$, and this gives $t_{first} = 0.667$, which is the same result that would be obtained without the central baffle. Suppose now that there are two s of equal length. An improvement to $t_{first} = 0.772$ can be achieved by placing the baffles at off-center locations in the two stages. Figure 5 shows results for systems up to 20 s. Table 1 gives the corresponding baffle locations. The optimization assumed each to be of the same length and ignored entrance effects at the transitions between stages, a justifiable assumption when the stage lengths are large. This assumption also means that the order of the various stages is unimportant.

Results reflecting entrance effects can be presumably be obtained from CFD codes, but caution is needed. Only recently has it been shown possible to generate a reasonably good washout function for diffusion free, laminar flow in a tube using Fluent (Tilton and Liu, 2008). Other postings on the Fluent website are inapplicable (e.g. for CSTR's where numerical diffusion would aid in approaching an exponential distribution) or show numerical diffusion large enough to grossly distort the first appearance time.

1	2	3	4	5	6	7	8	9	10	15	10
0.500	0.458	0.439	0.427	0.418	0.412	0.407	0.403	0.399	0.396	0.391	0.382
	0.542	0.500	0.478	0.463	0.452	0.444	0.437	0.432	0.427	0.407	0.404
		0.561	0.522	0.500	0.485	0.473	0.464	0.457	0.450	0.428	0.420
			0.573	0.537	0.515	0.500	0.488	0.479	0.471	0.445	0.438
				0.582	0.548	0.527	0.512	0.500	0.491	0.460	0.439
					0.588	0.556	0.536	0.521	0.509	0.474	0.454
						0.593	0.563	0.543	0.529	0.487	0.465
							0.597	0.568	0.550	0.501	0.476
								0.601	0.573	0.514	0.486
									0.604	0.527	0.496
										0.541	0.507
										0.556	0.517
										0.574	0.526
										0.591	0.537
										0.608	0.549
											0.560
											0.575
											0.587
											0.587
											0.615
0.667	0.712	0.735	0.751	0.763	0.771	0.778	0.783	0.787	0.793	0.800	0.807

Table 1. Optimal baffle positions for multi-stage, two-channel slit flow

The results to this point could conceivably be realized in real hardware and show a significant improvement in first appearance time. We now what appears to be an upper limit on the first appearance time in multi-stage laminar flow. We postulate a multi-channel device that can produce any desired reorientation of the fluid at the transition between stages. For a two-stage device, the optimal transformation appears to be complete flow inversion (Nauman, 1977) where the fluid is literally turned inside out so that what was nearest the wall in the first stage is at the centerline in the second stage and conversely. A conceptual design even exists for this (Nauman, 1977). A more complex transformation is needed between second and third stages, but the effects of the assumed-to-be-optimal transformation are easily simulated. The various streamlines emerging from the second stage are sorted according to residence time. This with the shortest residence times after the second stage will be placed near the wall in the third stage and conversely. The upper curve in Figure 5 shows the results. It is speculated that these represent an upper limit for first appearance times in multi-stage. The coiled tubes of Saxena and Nigam (1984) approach these numerical values but require the equivalent of many more stages.

Conclusions

The paper has discussed two problems which represent non-conventional goals for static mixers. The approach has been to select an objective function, either the entire washout function or the first appearance time of that function. Then a static mixer configuration was specified, and the design parameters of the configuration were optimized to using the simple random technique. This techniques is delightfully easy to program but computationally inefficient. They the fluid mechanics in the present examples were simple and could be quickly solved to any desired accuracy. Because of modern computing speeds, the approach has been used on much more complicated systems. For example, it might be applied to the functional optimization of wall temperatures in a heat exchanger in order to minimize the breadth of the thermal time distribution. The counterpart of this study applied to two-channel flows in a tube should also be feasible. The random techniques is unsuitable for full-blown CFD studies but can provide guidance for selecting parameters in detailed case studies.

References – Pertinent but not all are cited in the body

Saxena, A. K., and Nigam, K. D. P., "Coiled Configeration for Flow Inversion and Its Effect on Residence Time Distribution", *AIChE J.*, **30**, 363-368 (1984).

Mridha, Monisha; Nigam, K.D.P. Coiled flow inverter as an inline mixer, *Chemical Engineering Science*, 63, 1724-1732 (2008).

Smith, O. J., IV, Graham, D.R., and Palamara, J. E., Temporal Mixing, *AIChEJ* (2006), 52, 1780-1789.

Nauman, E. Bruce and Smith, Oliver J., IV, "Novel Static Mixers for Surge Dampening," Accepted for *Chem Eng Res Des*, 2008.

James N. Tilton, James N. and Liu, Minye "Spatial Age Distributions from CFD Computed Velocity Fields" ISMIP 6, August 2008, Niagara-on-the-Lake, Canada

Ring, T., Wang, Bin: Sang-Choi, Byung; Dhanasekharan, K. "Predicting Residence Time Distribution with Fluent," http://www.fluent.com/about/news/newsletters/04v13i1/a23.htm

Singh, Rajeev Kumar, "Tutorial guide of residence time" . http://www.cfd-online.com/Forum/fluent_archive_2006.cgi/read/40759

Nauman, E. B., "On Residence Time and Trajectory Calculations in Motionless Mixers", *Chem. Eng. J., 47,* 141 - 148 (1991).

Nauman, E. B.; Kothari, D.; Nigam, K. D. P., "Static Mixers to Promote Axial Mixing," *Chem. Eng. Res. Des,* 80, 1-5 (2002)

Nauman, E. Bruce, "Residence Time Theory," Invited 100th Anniversary Review for Industrial & Engineering Chemistry, *I&EC Research*, 47, 3752-3766 (2008)

Nauman, E. B. *Chemical Reactor Design, Optimization and Scaleup*, 2nd Edition, Wiley, New york, 2008

Nauman, E. B., "Nonisothermal Reactors: Theory and Application of Thermal Time Distributions", *Chem. Eng.Sci., 32,* 359-367 (1977).

Bhamidipati, S.; Singh, R. K. Thermal time distributions in tubular heat exchangers during aseptic processing of fluid foods. *Biotechnology Progress* (1994), 10(2), 230-6.\

M. R.; Hofmann, R.; Andrews, R. C. Case study comparisons of computational fluid dynamics (CFD) modeling versus tracer testing for determining clearwell residence times in drinking water treatment. Templeton, *Journal of Environmental Engineering and Science* (2006), 5(6), 529-536.

Haggerty, Roy; Wondzell, Steven M.; and Johnson, Matthew A. Power-law residence time distribution in the hyporheic stage of a 2nd-order mountain stream *Geophysical Research Letters* (2002), 29(13), 18-1 to 18-4.

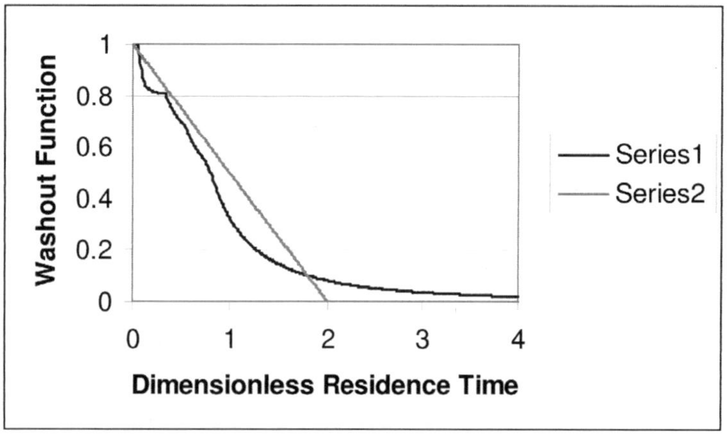

Figure 1. Five tube approximation to a flat RTD

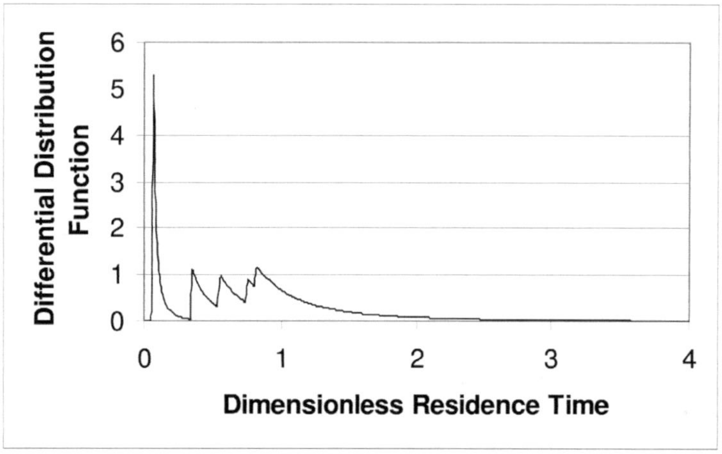

Figure 2. Density function, $f(t)$, for the five tube approximation

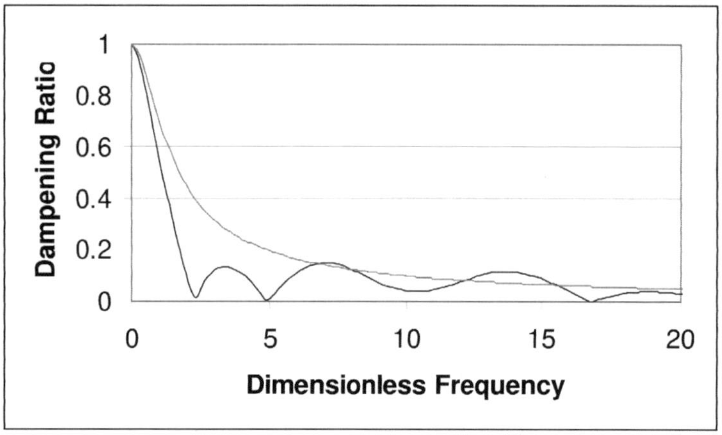

Figure 3. Response of the five tube approximation compared to that of a CSTR

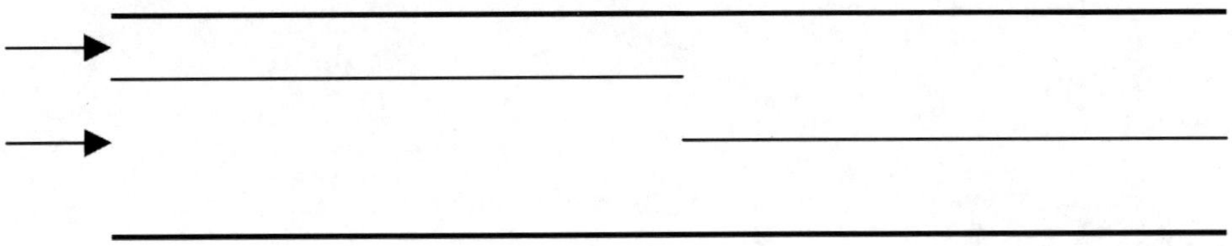

Figure 4. Two stages of the two-channel slit flow geometry

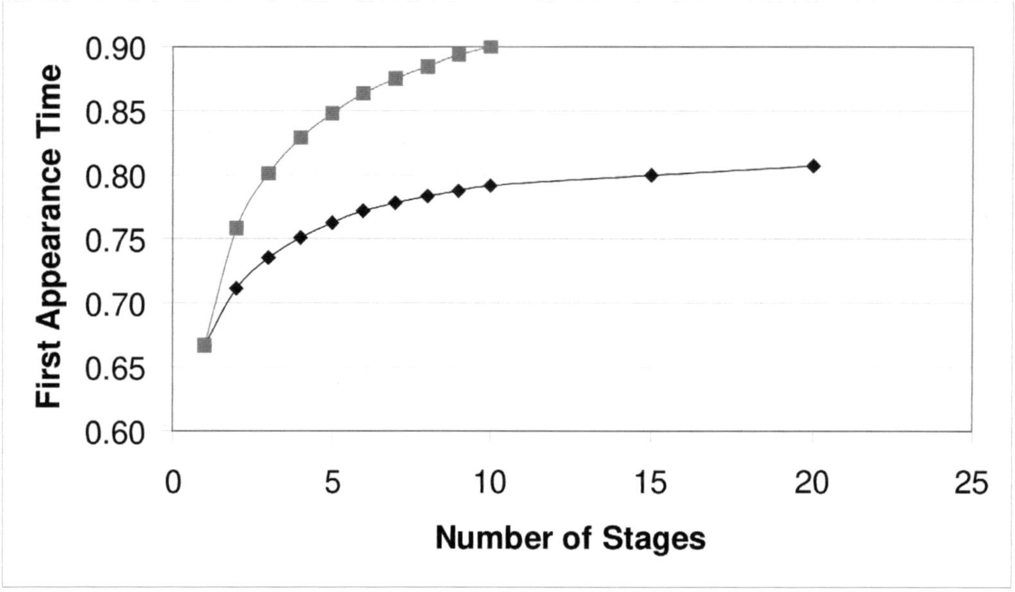

Figure 5. Optimized first appearance times for multi-stage, two-channel slit flow. The higher curve is speculated to be the upper limit possible in multi-stage slit flow.

495b Computational Characterization of Continuous Mixers Using Discrete Element Modeling Techniques

Atul Dubey and Fernando J. Muzzio. Chemical and Biochemical Engineering, Rutgers University, 98 Brett Road, Piscataway, NJ 08854

Granular mixing is an important process in a multitude of areas including but not limited to pharmaceuticals, food science, cosmetics manufacturing, and geological studies. In the pharmaceutical field, there is an incentive to move from batch mixing towards continuous mixing. Continuous mixing offers the advantage of added process control and manipulation, faster processing times and better control on the end product's content uniformity. Lately there has been a lot of emphasis on characterizing continuous blenders under specific operating conditions. However, computational methods have not been widely utilized to study continuous mixers. Discrete Element Modeling (DEM) offers a way to study and analyze the performance of such mixers. With the availability of increased computational power and commercial programs, it is now possible to quickly simulate and analyze many cases with varying test conditions. This study reports computational characterization of two commercial continuous blenders under a set of operating conditions defined by blade geometries and rotational speeds, blender processing angles, and the degree of cohesion in the ingredients.

495c Simulation of Flow Field and Circulation Time Phenomena In a Batch Rotor-Stator Mixer

Meng Yang, Karl R. Kevala, and **Richard V. Calabrese**. Chemical & Biomolecular Engineering, University of Maryland, Building 090, Room 2113, College Park, MD 20742-2111

Batch rotor-stator mixers are widely employed on the bench scale in the chemical and pharmaceutical process industries to demonstrate a variety of emulsion and dispersion processes, that must be scaled to geometrically or dynamically (continuous rather than batch) different equipment on the pilot scale. When the process fluid dynamics plays a significant role, a reasonable approach is to simulate both the bench and pilot scale flow field and deformation rate phenomena and match these on both scales to extrapolate bench scale data with respect to the number of mill head passes and shear rate history.

To this end, we are simulating the flow field in a Silverson L4R batch rotor-stator mixer with the milling head located off-axis in a 2 liter vessel. Since water is the working fluid, the flow is turbulent. We use a RANS formalism with a k - epsilon turbulence model and wall functions. Due to the importance of the rotor-stator interaction, a sliding mesh CFD technique must be employed. While time periodic convergence is relatively rapid close to the mill head, the pumping capacity is low. As a result, it is not possible with current computational resources to compute until a fully converged solution is reached throughout the entire tank. This is of little consequence if the goal is to understand rotor-stator interactions. However, it is a major barrier to allowing one way coupled particle tracking throughout the vessel, to determine the shear history and circulation time distribution of tracked particles.

At a moderate distance from the mixing head the effect of rotor-stator interactions is not felt, so the flow appears steady in time. A time periodic region, centered on the mill head can be defined by interrogating a sliding mesh solution for the entire tank. Beyond this region, the flow field is well described by an MRF simulation technique. As a result, the following hybrid technique can be used to significantly reduce the computational expense of particle tracking simulations. Based on mixing head geometry/ placement, vessel size/geometry, rotor speed and physical properties, the largest possible time periodic region can be defined. We then begin with an accurate, fully converged MRF solution, which always holds beyond the time periodic (outer) region and provides boundary conditions for a sliding mesh solution within the time periodic (inner) region. The sliding mesh algorithm is then begun for the inner region. This inner region is relatively small in volume, so convergence is rapid. The inner and outer solutions are then super imposed to provide a seamless time dependent velocity field for particle tracking trials. A single, repeatable period (in this case 90 degrees due to 4 rotor blades) of this simulation can be continually repeated as input to a fast particle tracking algorithm. Our previously reported computationally efficient tracking algorithm, which accounts for all of the forces acting on particles, can then calculate the trajectories of thousands of particles for numerous impeller revolutions to yield statistics for circulation time distribution and shear rate history.

We will demonstrate the hybrid velocity field CFD technique and particle tracking algorithm, and report the results of our simulations for several different cases. We will discuss our future plans in the context of our progress to date and what can realistically be accomplished in the short term.

495d Impeller Design for Optimal Performance In Tall Tanks

Minye Liu and Richard K. Grenville. DuPont Engineering Technology, DuPont Bldg. D5060, 1007 Market St., Wilmington, DE 19898

Tall tanks are often used in process industry for mixing. It is always important to design a set of impellers that achieve optimum mixing for the desired process result(s). One condition for efficient mixing is to generate a single flow loop in the whole tank in order to create effective top to bottom mixing. However, this is often a very challenging task to accomplish. Compartmentalized flow patterns in tall tanks are often observed where each of the impellers creates a mixing zone of it own. In order to design a set of impellers that work coordinately, many factors have to be considered, such as impeller type and diameter, distance between impellers, baffle sizes, etc. In this talk, we will discuss our CFD results of flow fields and residence time distributions in a tall tank with up to five impellers and show under what condition we achieve a single loop mixing in the tank.

495e The Use of Discrete Element Modeling (DEM) and Computational Fluid Dynamics (CFD) for Analysis of Solid-Fluid Mixing

Richard D. LaRoche and Wai Sam Wong. DEM Solutions, 24 Hanover Street, Suite 10, Lebanon, NH 03766

The use of computer aided engineering (CAE) tools for mixing analysis became mainstream with the advances in computational fluid dynamics (CFD) in the mid-1990's. CFD packages such as CFX, STAR-CD and FLUENT all introduced software capable of modelling a key mixing unit operation in the process industries – the stirred tank reactor. Progress over the last decade has led the validated use of CFD in modeling mixing in stirred tank reactors. CFD analysis is routinely used for determining 3D velocity fields, impeller power numbers and blendtime. The richness of this 3D analysis now gives engineers to generate probability statistics for turbulent energy dissipation rate and strain rate which are very useful complements to current scale-up methods in industrial fluid mixing practice.

CFD in stirred tank mixing analysis still faces some challenges in modeling multiphase mixing in stirred tanks using Eulerian-Eulerian CFD methods. For example, it has been problematic for researchers to predict the solids concentration profile in a solid-liquid stirred tank mixing application. Some researchers have had some success in adjusting multiphase parameters to match experimental data at a single scale -- the "belly curve" of solids concentration in a stirred tank as you traverse from bottom to top. But fixing these parameters has not led to a general ability to predict solids concentrations at different scales.

The author will outline an enhanced approach where the particle nature of the solids can be introduced through the use of Discrete Element Modelling (DEM) and coupled with CFD analysis. DEM is a numerical method which models the movement and contact between each particle. A DEM model particle can represent either a single particle or group of particles in the physical system. The most common and physically accurate implementation of DEM is the 'soft contact' approach which calculates the forces acting on each particle using models which can account for the particle mechanical and surface properties.

DEM is very well established as a powerful tool for studying the mechanics of granular bulks. Advanced DEM software tools such as EDEM can model both particles and equipment which enables direct application to complex processes. DEM provides a framework for investigating the relative effect of particle material properties, environmental conditions and equipment design. Factors such as particle shape, surface properties (e.g. cohesion and electrostatic charge), temperature and moisture content can be accounted for using appropriate models. DEM coupled with CFD is explored as an avenue to incorporate more of the physics necessary to allow multiphase CFD to break through into solid-liquid mixing analysis.

495f Interfacial Area Concentration Distribution Computed from An Interfacial Energy Balance

Richard L. Long, New Mexico State University, Chemical Engineering Department, P.O. Box 30001, MSC 3805, Las Cruces, NM 88003-8001

In fluid-fluid multiphase flow problems, the ensemble averaged interfacial area concentration balance equation can have a source term coming from the interfacial interactions between the phases. This type of balance can develop its source term from an interfacial energy balance. The primary competing approach is to partition the dispersed phase into a collection of size classes and develop source terms from breakage and coalescence terms that depend primarily on the energy dissipation rate per unit mass and the size class. The advantage of this approach is that the result is both the distribution of bubble sizes and the ensemble averaged bubble size. The disadvantage is the large number of popuation balance equations and source terms needed to make the computation. The current approach gives only an ensemble averaged interfacial area concentration, but it only requires a single additional scalar balance equation in addition to the usual formaulation of the two fluid model. Frequently, the ensemble averaged interfacial area concentration distribution is sufficient for multiphase reaction and mixing.

AUTHOR INDEX

AUTHOR INDEX

American Institute of Chemical Engineers
3 Park Avenue
New York, New York 10016-5991

Hydrogen Production for a Hydrogen Economy

Topical Conference at the 2009 AIChE Annual Meeting

Nashville, Tennessee, USA
8-13 November 2009

ISBN: 978-1-61567-917-1

AIChE